ISBN 978-1-331-12793-2
PIBN 10025853

This book is a reproduction of an important historical work. Forgotten Books uses
state-of-the-art technology to digitally reconstruct the work, preserving the original format
whilst repairing imperfections present in the aged copy. In rare cases, an imperfection in
the original, such as a blemish or missing page, may be replicated in our edition. We do,
however, repair the vast majority of imperfections successfully; any imperfections that
remain are intentionally left to preserve the state of such historical works.

# 1 MONTH OF
# FREE
# READING

## at
## www.ForgottenBooks.com

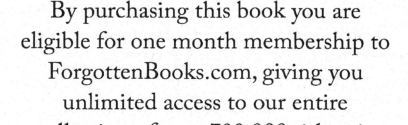

By purchasing this book you are eligible for one month membership to ForgottenBooks.com, giving you unlimited access to our entire collection of over 700,000 titles via our web site and mobile apps.

To claim your free month visit:
www.forgottenbooks.com/free25853

English
Français
Deutsche
Italiano
Español
Português

# www.forgottenbooks.com

**Mythology** Photography **Fiction**
Fishing Christianity **Art** Cooking
Essays Buddhism Freemasonry
Medicine **Biology** Music **Ancient
Egypt** Evolution Carpentry Physics
Dance Geology **Mathematics** Fitness
Shakespeare **Folklore** Yoga Marketing
**Confidence** Immortality Biographies
Poetry **Psychology** Witchcraft
Electronics Chemistry History **Law**
Accounting **Philosophy** Anthropology
Alchemy Drama Quantum Mechanics
Atheism Sexual Health **Ancient History**
**Entrepreneurship** Languages Sport
Paleontology Needlework Islam
**Metaphysics** Investment Archaeology
Parenting Statistics Criminology
**Motivational**

THIS BOOK IS DUE ON THE DATE
INDICATED BELOW AND IS SUB-
JECT TO AN OVERDUE FINE AS
POSTED AT THE CIRCULATION
DESK.

# THE UNIVERSITY OF CHICAGO
# SCIENCE SERIES

———

# THE BIOLOGY OF TWINS
## (MAMMALS)

THE UNIVERSITY OF CHICAGO PRESS
CHICAGO, ILLINOIS

Agents

THE BAKER & TAYLOR COMPANY
NEW YORK

THE CUNNINGHAM CURTISS & WELCH COMPANY
LOS ANGELES

THE CAMBRIDGE UNIVERSITY PRESS
LONDON AND EDINBURGH

THE MARUZEN-KABUSHIKI-KAISHA
TOKYO, OSAKA, KYOTO, FUKUOKA, SENDAI

THE MISSION BOOK COMPANY
SHANGHAI

KARL W. HIERSEMANN

# THE
# BIOLOGY OF TWINS
## (MAMMALS)

*By*

HORATIO HACKETT NEWMAN

THE UNIVERSITY OF CHICAGO PRESS
CHICAGO, ILLINOIS

Composed and Printed By
The University of Chicago Press
·Chicago, Illinois, U.S.A.

# PREFACE

The present volume brings together for the first time a considerable mass of data dealing with the phenomenon of twins in man and other mammals. Twins are so inherently interesting to so many people that it is hoped by the writer that some light on how twins "happen" will be welcomed by the general reader as well as by the biologist. There are many thoroughly interesting and nontechnical phases of twin-biology that will appeal to anyone who is a twin or has personal acquaintance with twins. There must be certain other phases of the subject, however, that are largely of value to the professional biologist. It has been the aim of this book to satisfy both the general and the technical reader without sacrificing unduly the demands of simplicity on the one hand or of scientific adequacy on the other.

H. H. N.

# CONTENTS

# INTRODUCTION

Everyone is or should be interested in twins. It is my task to bring together the facts about twins and to show the bearings of these facts on fundamental problems of biology. It would be an easy task to write a treatise on twins for the expert embryologist or student of genetics, but it is far less easy to present this material adequately and to make it crystal-clear to those who are not specialists. It has seemed necessary to strike a compromise between a technical presentation and a somewhat popular treatment of the subject. Much of the more general matter in all of the chapters will be found available for the general reader, but some of the descriptive embryology, which is the foundation of our special knowledge, will be rather difficult even to the embryologist. Every effort has been made to simplify this part of the book without running the risk of denaturizing it. Again, there are parts of the chapters on heredity that will appeal especially to students of that important subject, but will have only a minor appeal to the general reader.

It is impossible to avoid technical terms, especially in descriptive embryology, but where certain simple terms serve as well as the more technical ones they will be used. In referring to early embryos of human beings or of armadillos we might use the words "blastodermic vesicle" or "blastocyst," but it will be simpler to use the common word "egg" for the early mammalian embryo and its membranes. Again, in speaking of

the sex of twins we shall avoid for more reasons than
one the terms "homosexual"[1] and "heterosexual,"
which refer respectively to twins both of the same sex
and twins of opposite sexes, and shall use the less
objectionable terms "same-sexed" and "opposite-
sexed."

Where the use of technical terms appears unavoid-
able, the reader will find that, as a rule, a term is defined
when first used and possibly in more places than one.
Frequently, too, when the verbal description is difficult
of comprehension the illustrations will give the essential
information.

In this book an attempt is made to gather from many
sources the facts about *mammalian*[2] twins and to unify
these varied situations into one point of view. My
own interest in this subject has grown out of eight
years' study of what is perhaps the most striking case
of twin-production known: that exhibited by the nine-
banded armadillo of Texas (*Dasypus novemcinctus*).
The somewhat disproportionate space devoted to the
phenomenon of polyembryony in this species needs no
apology. The various aspects of its biology have been
more extensively studied than those of any other
species, and an author may be forgiven for emphasizing
the parts of his subject with which he has a first-hand
acquaintance.

[1] The term "homosexual" is extensively used in the literature
dealing with abnormal sex relations and is therefore pre-empted.

[2] An extensive chapter on twinning among the various vertebrate
classes below the mammals would appear to lend completeness to this
volume, but a review of the extensive literature convinces me that such
a chapter would be more confusing than helpful to the general reader.
I shall therefore deal with twinning in mammals only.

No simple definition of *twinning* can be given, for the term has come to be applied to at least four distinct situations:

1. The production from plural eggs (zygotes) of plural offspring in species that habitually produce but a single offspring at a birth. This would include some twins in man, as well as triplets, quadruplets, and larger sets of simultaneously born offspring. Cattle and sheep twins and triplets also belong to this category.

2. The production of plural offspring either sporadically or as a specific character, from a single fertilized egg (zygote). Such twins, quadruplets, or larger sets of offspring are known as *monozygotic*, and this mode of reproduction is known as *polyembryony*. Two species of armadillo belonging to the genus *Dasypus* exhibit specific polyembryony; also there appears to be sporadic polyembryony in man and possibly in other species.

3. The habitual or specific production of paired offspring, each of which is derived from one fertilized egg. Such *dizygotic* twins are specific for the armadillo *Euphractus villosus*, and possibly for other species of that genus.

4. The sporadic production of conjoined twins and various types of double monsters. This is the category to which Siamese twins and similar monstrosities belong; the condition is known to occur in many mammalian species, although very little study of the phenomenon has been made except for man. Some conjoined twins are evidently monozygotic and are due to processes akin to polyembryony, but others are evidently due to a secondary fusion of dizygotic individuals.

The one fact that stands out above all others in this connection is that there are really but two distinct kinds of twins: those that come from a single egg and those that come from two or more eggs. The former type involves some process of fission or division of a germ, which is at first a single individual, into two or more complete or, in certain cases of incomplete twinning, partial individuals. This is really the only *true* type of twinning.

When, however, twins or triplets, etc., are derived from two or more eggs, the biology of the situation differs very little from the ordinary phenomenon of multiple births, as seen in swine, dogs, cats, and other common mammals. In these animals we do not use the term twin, triplet, or quadruplet, because this giving birth to a number of offspring at a time is the normal condition, whereas a single offspring is exceptional. It is only in those species that normally give birth to but one offspring at a time that we note especially the more or less frequent exceptions to the rule, and refer to double births as twins, triple births as triplets, quadruple births as quadruplets, etc.

Monozygotic twinning, where a single egg produces plural offspring, is therefore a phenomenon that should be considered as only a phase of the much more general phenomenon of symmetrical division. The development of the right- and left-hand homologous organs in a bilateral organism is essentially a twinning process, for it involves the division of a median unpaired primordium into two equivalent parts, one of which is the mirror-image of the other. In many cases this twinning of parts may be more or less completely inhibited, so that a failure of certain parts to divide occurs and a single median structure appears. The paired eyes, for example, may fail to develop and a single median cyclopic eye may result. Just as there may be an inhibition of the normal culmination of the process of bilateral division, so there is frequently an excess of division resulting in two bilateral structures becoming completely separated, as when a single individual develops two heads or two tails, while the remainder is a more or less normal individual. The whole matter of bilateral development appears to be quantitative in nature, in that the same type of process may go not so far or farther than normal.

There are known for man as well as for the lower vertebrates all stages of twinning, ranging from incompletely bilateral forms that are subnormal, such as cyclopic monsters, through various grades of supernormal forms, such as two-headed types, double monsters, and conjoined twins, culminating in completely separate monozygotic or duplicate twins. That all these types are merely phases of the same process of *bilateral doubling* is, I believe, beyond question, as will be shown in the chapters that follow.

The phenomenon of twinning then will be seen to be a very fundamental process, one almost universal in the field of biology. *For wherever we have bilateral doubling we have twinning in some form.*

All expressions of the twinning process involve the same biological problems: those of symmetry, heredity, and sex; but perhaps the process of duplicate twinning proper, where two or more completely separate individuals are produced, affords the most available material for a study of these problems.

The problems of twinning and those of sex and of heredity are inextricably interwoven; each of them appears to be a corollary of the other. It is not surprising then that some of the most conclusive evidence as to the nature and mode of sex-determination and much new light on the nature and limits of hereditary control come from a study of twins.

A collection of data upon twinning in mammals brings together more biological curiosities than is furnished by any other field of similar scope with which I am acquainted. The armadillos furnish two strangely unique situations quite opposite in character. In *Dasypus* there is the splitting up of a single egg into a number of separate embryos, while in *Euphractus* two originally separate eggs secondarily undergo extensive fusion of their membranes so as to produce monochorial[1] twins quite deceptively like those known to be monozygotic. Twinning in cattle involves the very odd phenomenon of the *freemartin*, where usually a female born co-twin to a male is sterile and shows certain male characteristics. An analysis of this peculiar situation goes far toward clearing up the problems of the nature of sex and of the factors which control it.

There are also many strange facts about the various kinds of human twins. Duplicate or identical twins are of unusual interest; conjoined twins of the

[1] Monochorial twins are surrounded by a single chorionic membrane. Usually a single chorion covers but one embryo and comes from a single egg; hence it is customarily assumed that the monochorial condition implies a monozygotic origin.

Siamese type and so-called "parasite" twins are also oddities that have long been known but little understood. Owing to the special interest that naturally attaches to human affairs, the first chapter will be devoted to human twins, although so little is known about their mode of development. The second chapter gives a full account of the process of polyembryonic twinning in the armadillo, *Dasypus novemcinctus*, and there is reason to believe that in this material we have the only key to the mechanics of human twinning.

I wish to express my thanks to my friends W. H. Osgoode and F. R. Lillie for reading the manuscript and for useful suggestions; to Mr. Kenji Toda for his aid in preparing illustrations; and to those authors from whom figures have been borrowed.

# CHAPTER I

## VARIOUS KINDS OF HUMAN TWINS

Human twins have always been objects of especial interest, partly perhaps on account of the fine humor of the situation, and partly because of the frequent occurrence of "look-alike" twins or "duplicate" twins. Popular interest has quite naturally focused upon the striking similarity, amounting in some cases to almost complete identity, which exists in certain types of twins; this emphasis upon resemblance is justified by the biological analysis of twinning, as is shown in what follows.

Biologists have for some time recognized at least two distinct types of human twins: *fraternal* and *duplicate*. *Fraternal* twins may or may not be same-sexed, are usually no more alike than are brothers and sisters, and are believed to be dizygotic, derived from two fertilized eggs. *Duplicate* or identical twins are always of the same sex, are almost identical, and are believed to be monozygotic, derived from a single fertilized egg.

No twins occur in nature at all comparable with the old favorite type which has done such signal service in the drama and in fiction (not to mention the "movies"), wherein brother and sister are identical. Science, however, can afford to be magnanimous, so let us as a concession to art include in our list these "literary twins."

Another type of human twins, stranger than fiction, is found in conjoined twins and double monsters, of which the Siamese twins furnish a stock example. There appears to be a graded series of types ranging between duplicate twins and the less monstrous types of conjoined twins; similarly, the conjoined twins appear to grade into the various kinds of double monsters. Some conjoined twins are very lightly connected and some double monsters are so closely united that one individual may be a mere degenerate parasite upon the other. The fact that very lightly conjoined twins exist would point to the probability that such twins might sometimes be born separately. This is Wilder's view of the relation between duplicate twins and double monsters. Lightly conjoined twins are always of the same sex and are strikingly similar; there seems to be no question as to their monozygotic origin. What more natural, therefore, than to infer that separate twins which are of the same sex and strikingly alike are also *monozygotic?*

Considerable direct and indirect evidence that monozygotic twins are of frequent occurrence in man is available. Perhaps the most conclusive evidence for this idea is to be found in a study of the sex-ratios of twins. Data are available from several different sources, but perhaps the best is that given by Nichols,[1] which is herewith presented:

|  | Sex of Twins | | |
|---|---|---|---|
|  | ♂♂ | ♂♀ | ♀♀ |
| Frequency of occurrence.... | 234,497 | 264,098 | 219,312 |

[1] J. B. Nichols, *Memoirs of the American Anthropological Association*, I (1907).

This is approximately a ratio of $\frac{\male\male}{1}$ : $\frac{\male\female}{1}$ $\frac{\female\female}{1}$. Now if all twin births in man are dizygotic and the sex is predetermined at the time of fertilization, as there is every reason to believe, the sex-ratios of twins should be: $\frac{\male\male}{1}$ : $\frac{\male\female}{2}$ : $\frac{\female\female}{1}$. There are, however, nearly twice as many same-sexed twins as there should be on this basis, and the only satisfactory explanation of this discrepancy between the observed and the expected ratios appears to be that *nearly half of all same-sexed twins are monozygotic and hence morphologically stand for but one individual to the pair*. It would appear then that in Nichols' data the difference between the actual numbers of same-sexed twins and half the number of opposite-sexed twins will give the probable number of monozygotic twins of each sex; this would be 102,448 monozygotic male twins and 87,263 monozygotic female twins. About one-fourth of all human twins then, if this reasoning holds, are monozygotic. My own observation and that of biologists with whom I have discussed this point agree very closely with this conclusion.

The form of the human uterus and the intra-uterine relations of twins serve as another line of evidence favoring the existence of monozygotic human twins. The human uterus is of the simple type, resembling that of the armadillo *Dasypus*, and is not adapted for ordinary multiple gestation. Such a uterus, however, is apparently as favorable for polyembryony (the production of plural embryos from one egg) as that of our armadillos, in which this kind of reproduction occurs normally.

The evidence furnished by the data collected by obstetricians of twins *in utero* also favors the existence of monozygotic twins. Frequently this evidence is lacking in very essential points. Sometimes, for example, the sex is not mentioned, and never, so far as I am aware, is there information about the number of corpora lutea present. The importance of the latter data cannot be overemphasized in this connection; a knowledge of whether one or more corpora lutea are present would furnish a crucial test of the number of ova concerned in a given pregnancy. In not a single case of human multiple births, so far as I am aware, has the number of corpora lutea been noted. The reason for this is obvious: to secure this information would require an operation during pregnancy. It seems quite probable, however, that post-mortem examinations of pregnant women, and of those dying after abdominal operations or during parturition, might serve to give occasionally just the kind of information that we must have in order to prove the existence of monozygotic twinning in man. I would therefore urge upon surgeons and obstetricians the importance of collecting information as to the corpora lutea whenever opportunity arises.

Although data as to intra-uterine relations are incomplete and inconclusive, they form the only really direct evidence now available on the mode of twinning in man. The most reliable collection of such data is that of O. Schultze.[1] He grouped his types into four categories which have been given by Wilder,[2] together with the latter's comments, as follows:

*Case I.*—Two separate blastodermic vesicles with two deciduous reflexae and two placentae. This case is probably one in

[1] O. Schultze, Leipzig, 1897.
[2] H. H. Wilder, *American Journal of Anatomy*, III (1904).

which there are two separate eggs, either from the same or from opposite oviducts, and implanted at some little distance from one another. In one case investigated by Von Kölliker, the two deciduae were distinct but partially adherent over the surfaces in mutual contact, and in another the contact surfaces had fused into a single wall into which, from the two opposite sides, the chorionic villi of the two embryos had grown. In addition to this, one of the placentae was of the type known as placenta marginata, caused by a fold of the decidua. (This is evidently a normal multiple birth, a condition hard to accomplish in a uterus of the shape found in human beings, and often attended by such phenomena as adhesions, fusions, and foldings, all indicative of crowding and nothing else.—Wilder.)

*Case II.*—Two separate blastodermic vesicles inclosed in a single decidua. Placentae fused with one another but with two separate sets of umbilical vessels. Two chorions fused at the point of contact. This case is more frequent than (I) but apparently results from the same general cause, i.e., two separate eggs, which are, however, implanted nearer together. This would seem more likely to happen if both eggs came from the same side. (The conditions are seen to be similar to those of (I), the greater degree of fusion being well accounted for by the approximation of the two eggs to one another.—Wilder.)

*Case III.*—Two amnions and two umbilical cords with a single placenta in the middle of which the two cords meet and upon which the umbilical vessels closely anastomose. These are inclosed in a single chorion and covered with a single decidua reflexa. This case is said by Hyrtl to be more frequent than (I) and (II), but not as frequent, according to Späth. The twins are always of the same sex. Schultze says that the explanation of this singular condition is *zweifelhaft* and gives the following possible explanations: (1) At first two chorions, as in (II), the contact wall between which becomes absorbed later; (2) may have come from a single egg with a double yolk, or (3) from an ovarial egg with two nuclei (cf. Franqué, 1898, Stoeckel, 1899, H. Rabl, 1899, and von Schuhmacher u. Schwarz, 1900). It is conceivable that from such an egg as this last, two blastodermic vesicles and two chorions could develop within one zona pellucida, at a later stage of which two chorions could fuse.  Von Kölliker considers it more probable, however, that in such a case the egg would develop two embryonic areas upon a single blastodermic vesicle and that a single chorion would then be the natural result. Each embryonic area would develop its own amnion.  In this case the two allantois would necessarily fuse, being included in a single chorion, and there would come to be between the two embryos a single (common) yolk sac with two yolk stalks.  Von Kölliker has observed such cases in hen's eggs (but without fusion of the allantoids).  M. Braun has seen it in lizards, and Panum describes separate embryonal areas upon one yolk (hen's egg). See also Koestner's figure of a double egg of *Pristiurus*, 1898. (This case seems to put us on the right track regarding the origin of duplicate twins, especially since it is stated that the twins are always of the same sex, and although observations of later physical identity are wanting, it seems safe to assume it.  It would seem highly improbable, however, that duplicate twins would arise from an ovarian egg with two nuclei, since in such case the fertilization could be effected only by means of two spermatozoa, thus introducing two paternal characters; but if we reject all of Schultze's alternatives and substitute the possibility suggested above, that of the complete separation of the two blastomeres resulting from the first cleavage of the fertilized egg, the two components would still remain within one zona pellucida and would later become inclosed within a single chorion, which would develop a single placenta to which each allantois would become later attached.  Each blastomere would undoubtedly form at first an independent blastodermic vesicle, but the close association of the two would readily tend toward a fusion of the contact surfaces, thus forming a single vesicle upon the surface of which are two embryonic areas.  If far enough apart from one another, each would develop its own amnion, but if near together a common amnion would result, thus producing the condition given in Case IV.  The whole matter of the actual condition of the development of the two associated embryos is very obscure, as there are but scattered and insufficient data bearing upon the case.  It will receive a more extended consideration later on, under the heading "Origin of Composite Monsters" and

"Other Recent Theories Concerning the Genesis of Composite Monsters."—Wilder.)

*Case IV.*—Similar to (III), but with both embryos inclosed in a single amnion. This is a very rare case, explicable only by postulating a single blastodermic vesicle upon which the two embryonal areas are nearly or entirely in contact with one another, a case which has been described by several authors as occurring in the hen's egg. In such a case there would be an almost irresistible tendency toward the fusion of the two embryos along the line of mutual contact, thus producing some form of composite monster. [Schultze says *Doppelmissbildungen*, but I use the word *double* in a more restricted sense, as explained below.]

(As Case II is seen to be a variation of Case I with the two embryos nearer together, so Case IV is seen to be a similar variation of Case III, with a similar result, i.e., the more complete fusion of the parts, although here, owing to the direct connection of the two embryos, the fusion is liable to extend also to these and produce abnormal results. There are thus primarily not four, but two cases, corresponding to the two types of twins, fraternal and duplicate. The close connection of IV and III suggests what may have already occurred to the reader: that many cases of compound monsters come under the same category as separate duplicates. This is quite probable, but such forms, arising from a secondary fusion, would be more asymmetrical and more or less unequal, and would come under the class of autosite and parasite rather than that of symmetrical, or genuine double monsters.—Wilder.)

This rather extensive quotation from Wilder's monograph on *Duplicate Twins and Double Monsters* is given largely to show the inadequacy of the embryo logical data upon which conclusions as to the mode of origin of twins are based. In addition, one will readily gain the impression that there is a very wide diversity of interpretations of the facts, based largely upon assumed analogies with conditions found in other vertebrates and even invertebrates. About the only conclusion which is actually warranted by the facts is that there are two kinds of human twins, *fraternal* (dizygotie) and *duplicate* (monozygotic). There is such a diversity of conditions, however, that it is impossible in some cases to decide whether a given set belongs to one or the other category. This situation is in marked contrast with that seen in the armadillos (chap. ii), where the intra-uterine relations are practically uniform in all pregnancies, indicating that the peculiar mechanism which brings about twinning is rigidly standardized. In human twins, however, twinning is by no means a single, fixed process, but is highly variable, evidently beginning earlier and being more complete in some cases than in others. Wilder is probably correct in considering that the various symmetrical types of double monsters belong to the same series as separate duplicate twins. I would suggest that they are merely more or less incompletely separated monozygotic twins, in which the twinning process begins later than in the separate twins and then fails to be fully carried out.

Further evidence that duplicate twins are monozygotic is derived from a study of identity between twins and certain cases of symmetry reversal seen between them. These matters will be taken up in a subsequent chapter.

There is, as we have seen, some embryological evidence derived from the fetal membranes of certain twins that they are monozygotic. There are also certain twins strikingly identical. Unfortunately, however, there is no case in which both kinds of data have been secured for the same sets of twins. This is the information which, together with the data on the corpora lutea, would be necessary to raise the phenomenon of monozygotic twinning in man from the realm of probability to that of demonstrated fact. Even if these data were obtained, we should still be far from a real understanding of the mechanics of twinning in man.

### THEORIES OF THE MODE OF TWINNING IN MAN

Assuming that duplicate twins in man are monozygotic, how is twinning accomplished? In the quotation of Schultze and from Wilder's comments we may glean a knowledge of practically all the various hypotheses on this subject. Some of these are scarcely of sufficient importance to deserve notice. Origin from a double-yolked egg, for example, would scarcely be possible in a mammal. Wilder rightly excludes the possible origin of twins from binucleated eggs, because there would have to be two sperms, and this would introduce a degree of diversity that does not occur. Von Kölliker's suggestion that a single egg may occasionally produce two embryonic areas which might or might not develop separate amnions, depending on the degree of juxtaposition of the two areas, is well worth consideration. It seems far more probable than Wilder's original theory of the complete separation of the two blastomeres resulting from the first cleavage division of a fertilized egg. Wilder has recently abandoned this extreme "blastotomy" theory under the influence of the discoveries of Newman and Patterson on the armadillo, and is inclined to agree with these writers in their suggestion that the origin of human duplicate twins probably resembles that of the armadillo quadruplets.

The theory that duplicate human twins arise by some sort of early fission process, probably initiated in the ectoderm, is rendered probable by the facts observed in the earliest known human embryos, notably those of Peters, of Frassi, and of Bryce and Teacher. It seems certain that the amnion in the human embryo is not a folding of the somatopleure, but arises as a hollow in a ball of ectoderm, as in the armadillo. This type of amnion formation would furnish the requisite mechanism for the production of ectodermic outgrowths from which, as in the armadillo *Dasypus*, the primordia of twins could take their origin.

Though abandoning the crude idea of "blastotomy" origin of duplicate twins, Wilder still clings to the idea that the hereditary differentiation between the twins is to be traced back to events taking place during the first cleavage. In this he agrees with my own position taken in connection with armadillo quadruplets.

My opinion regarding the mode of origin of human duplicate twins was stated in the concluding paragraph of my latest paper on "Heredity and Symmetry in Armadillo Quadruplets":[1]

I am inclined to believe that duplicate human twins become physiologically isolated at a considerably earlier period than do armadillo quadruplets, and my reason for this belief is founded on the fact that there is so little mirror-imaging in the former and so much in the latter. It appears to be a good general rule that the earlier the separation the more complete is the reorganization of symmetry relations in the separate individuals and the less residuum of the original common symmetry. Double monsters doubtless begin to separate comparatively late in ontogeny and hence (sometimes) show very pronounced mirror-imaging.

[1] H. H. Newman, *Biological Bulletin*, XXX, No. 2 (1916).

Since it seems entirely probable that human duplicate twins are separated at a much earlier period than are armadillo quadruplets, it may not be unreasonable to look for this separation at some period of cleavage. Or there may be a division of the inner-cell mass into the primordia of the two embryos. The problem of the exact origin of duplicate human twins is, however, likely to remain unsolved for a long time to come.

### CONJOINED TWINS AND DOUBLE MONSTERS

Double individuals have for a long time attracted the attention of obstetricians and embryologists, and there is an extensive literature on the subject, with little value, however, for a book of this sort. Certain of the lower-grade conjoined twins, the Siamese twins, for instance, have stimulated human curiosity to such an extent that they have been exploited as freaks in circus side shows and museums the world over. The late P. T. Barnum owed a considerable measure of his early success to his discovery and exhibition of the Siamese brethren. Every degree of junction between twins has been noted, ranging from mere fusion of parts of the skin that can be and have been cut apart without injury to extreme unions involving the head and entire trunk. An abbreviated classified list of types of composite monsters, based on Wilder's extensive data, will serve to show the range of forms included.

There are distinguished two main types: one "in which the components or component parts are equal to and the symmetrical equivalents of one another: *Diplopagi*"; the other, "unequal and asymmetrical monsters, one component of which is smaller than and dependent upon the other: *Autosite and Parasite.*"

1. *Diplopagi.*—These are subdivided into two groups characterized as follows: (*a*) "each of the two components complete or nearly so"; (*b*) "the two components equal to one another, but each one less than an entire individual." In the first group are included a considerable number of subtypes based on the point of juncture, some being joined at the head, others at the thorax, others at the sacrum. Names have been given to these types as follows: craniopagi, thoracopagi, and pygopagi. They may be back to back, side by side, or face to face. In fact, the point of juncture may be such that they are united so as to face at almost any angle. Yet it must be noted that they are not joined in any haphazard way, but are so placed that they form two bilaterally symmetrical objects. In fact, one is the mirror-image of the other, the plane of the imaginary mirror being that in which the junction exists. A number of types of *cosmobia* or symmetrical conjoined twins are shown in Figs. 1 and 2, and are described in the legends.

2. *Autosite and parasite.*—Sometimes the parasite is merely a head or a head and arms attached to the autosite at or near the epigastrium or upper part of the abdomen. At other times the parasite consists of legs and part of the lower body, without a head, attached as in the first case. Or, finally, the parasite may be a supernumerary head or face attached to the side or back of the head of the autosite.

Extreme conditions are those in which the parasite is within the autosite. These form tumors usually in the body cavity. They range from almost complete fetuses to mere masses of tissue sometimes containing

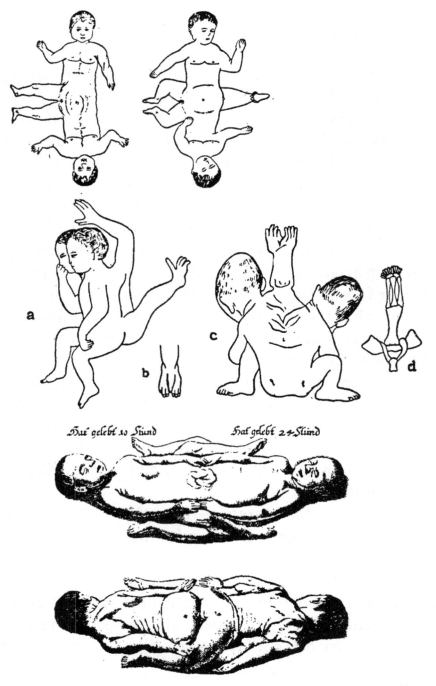

FIG. 1.—Various types of double monsters (from Wilder). These are all strongly conjoined and consist of less than two complete individuals. Note the inequality in size of the right-hand middle pair, the median, partially double arms and legs in several of the twins, and the symmetrical relations of the two individuals in all cases.

teeth or hair. One strange case of parasite and auto-site is cited by Blundell[1] and may be worth describing in detail. This is the case of "a boy who was literally

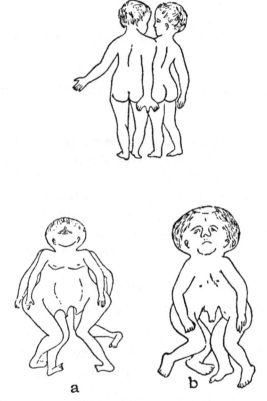

a
b

FIG. 2.—The upper figure shows Renault's twins (after Wilder), which approach the condition of the Siamese twins, where both individuals are complete or nearly so. The lower twins are typical Janus monsters: (a) a Cyclopian; (b) a case in which what appears to be a single broad face is really a double face in which the inner half of each component has been suppressed.

and without evasion with child, for the fetus was contained in a sac communicating with the abdomen and was connected to the side of the cyst by a short

[1] *London Lancet,* 1828–29, p. 260.

umbilical cord; nor did the fetus make its appearance until the boy was eight or ten years old, when, after much enlargement of pregnancy and subsequent flooding, the boy died."

### MODE OF ORIGIN OF DOUBLE MONSTERS

Wilder in his monograph of 1904 considers that double monsters are simply united or unseparated duplicate twins:

To one who sees in separate twins the result of the total separation of the first two blastomeres of a developing ovum there is but one rational explanation of diplopagi, or those composite monsters in which the two components are the duplicates of one another and symmetrically united, namely, *that here a similar tendency to separation has been left incomplete, causing doubling of those parts only in which the interrelations have been severed.*

Recently Wilder has abandoned his idea that duplicate twins and double monsters are derived by the complete or partial separation of the first two blastomeres, but adheres firmly to the idea that each individual traces back its cell lineage to one of those cells.

Various other theories involving the idea of fusion, as opposed to that of incomplete separation, have been advanced. Fischer as early as 1866 supposes that double monsters are the result of *an early total fission of the embryo followed by a secondary fusion of parts.* He says that they

are invariably the product of a single ovum, with a single vitellus and vitelline membrane, upon which a double cicatricula, or two primitive traces, are developed. The several forms of double malformation, the degree of duplicity, the character and extent of the fusion, all result from the proximity and relative position of the neural axes of the two more or less definite primitive traces developed on the vitelline membrane of a single ovum.

Considering how little was really known of embryology at the time, this idea of Fischer shows surprising insight. I am inclined to believe that this explanation comes very close to the true one. It will be remembered that in the account of intra-uterine relations of duplicate twins a condition was described in which the twins were not only monochorial but mon-amniotic (contained within a single amnion). This appears to me to present many possibilities for fusions. If we suppose that twins arise by some process of fission, in general like that in the armadillo, it is likely that in some cases the outgrowths arise so close together that only a single amnion is formed, and in such cases the visceral parts in particular would be likely to fuse or remain unseparated, as in the more pronounced types of diplopagi. Even in those cases that succeed in maintaining a disjunction until the body parts are completely separated it would be possible, or even highly probable, that, through the crowding incident upon growth within one amnion, surface fusions more or less extensive would occur. Wilder offers as an objection to this and other theories involving the idea of secondary fusion of originally separate primordia, that it is incompatible with the fact that the two components of a double monster are strictly bilaterally symmetrical and that "there is no force to oversee and adjust the two components in the exact relationship necessary for the result." This objection loses force, I believe, if the two components are viewed as outgrowths lying in a bilateral position on the germ and held in that position by their embryonic connections. It is hardly likely that two outgrowths could turn around or reverse their positions, at least not until the umbilical cords became elongated. Even armadillo quadruplets which are in separate amnia are found in advanced stages lying in positions such that, should fusions occur between adjacent individuals, they would unite to form diplopagi strictly bilaterally symmetrical with reference to one another.

It is highly probable that certain cases of doubling in head and posterior regions of human twins are due to disturbances in the process of concrescence of the right- and left-hand component of such embryonic primordia as the neural groove and the ventral body suture. In various vertebrates all sorts of partial doubling may be produced by experiment. For example, I have worked with certain strains of hybrid fish in which double-headed and double-tailed fish have been the result of abnormal developmental conditions. It may well be true, then, and probably is, that not all cases of doubling in human embryos are due to the same type of cause; some may be due to fission and others to fusion. Where we have so little evidence on the embryological side, it seems unwise to speculate further upon the causes of twinning and doubling in man.

The only real clue as to the mode of twinning in man comes from a study of polyembryonic development in the armadillo. For various reasons it is believed that the process of monozygotic twinning in man is essentially the same as that of the armadillo; hence a detailed study of the facts revealed by a study of the armadillo is the nearest approach possible today to a direct study of human twinning.

# CHAPTER II

## TWINNING (POLYEMBRYONY) IN THE ARMADILLO, *DASYPUS NOVEMCINCTUS*

### HISTORICAL

Polyembryony[1] is exhibited by two closely allied species of armadillo belonging to the genus *Dasypus*[2] (*Tatusia* or *Tatu* of some writers), *D. novemcinctus* and *D. hybridus*. Certain significant facts have been known about the latter species for about thirty years. H. von Jhering in 1885 and in 1886 published two brief notes in which he stated that he had secured two pregnant females of this species, the uterus of each of which contained eight fetuses, all inclosed within a single chorion. All in both sets were males. On the basis of this observation he published in a subsequent note the suggestion that all of the embryos of each set were a product of the splitting of a single fertilized egg into a number of separate embryonic primordia. Although the bearings of this situation on important problems of sex-determination and heredity were not appreciated by von Jhering, it must be acknowledged that, by a flash of insight, he arrived at an entirely correct diagnosis of the underlying biological significance of the observed conditions.

[1] Polyembryony is a unique mode of twinning in which plural offspring are derived from a single fertilized egg.

[2] Although most writers who have dealt with this species have given it the generic name *Tatusia*, the laws of priority favor the generic name *Dasypus*.

Further investigations concerning the mechanics of this unique type of reproduction were forestalled by the curiously misleading investigation of Rosner,[1] who in 1901 published an account of a microscopic study of the ovaries of a single specimen of *D. novemcinctus* sent him by von Jhering. Rosner found that many of the ripe follicles contained several eggs and that the two largest follicles each contained four eggs, a number corresponding to that of the fetuses of a litter. On the basis of these observations he decided that von Jhering's theory of the origin of the set of embryos from a single fertilized egg was incorrect, and that the four embryos typical for a litter in *D. novemcinctus* were derived from a nest of four eggs inclosed in a single follicle. No attempt was made to account for the fact that all in a litter were of the same sex. Unfortunately Rosner happened to examine a pair of pathological ovaries, as has been abundantly shown through my own studies and confirmed by the independent observations of several other investigators. It is very unusual to find more than a single egg in a follicle; in fact, only one pair of ovaries out of nearly thirty that have been sectioned for my studies shows any but normal monovic follicles. Since, as Rosner erroneously claimed, the four quadruplets came from four fused ova, the problem appeared to be solved. On this account, perhaps, no special interest was taken in the armadillo situation for some time.

In the year 1909 the question was reopened, when almost simultaneously there appeared preliminary accounts of the conditions in two species of *Dasypus*.

[1] M. A. Rosner, *Bull. Acad. Sci. de Cracovie*, 1901.

Fernandez[1] published an account of several embryonic stages of the *Mulita* (*D. hybridus*) taken in the Argentine Republic; and Newman and Patterson[2] published a preliminary report, based upon some advanced embryonic stages, of the conditions existing in *D. novemcinctus Texanus*—the Texas armadillo. Since the conditions in the two species are essentially similar and since much more has been published about the latter species, the main account of polyembryony in the armadillos will be based upon conditions worked out for the Texas species.

ECOLOGY AND HABITS OF THE TEXAS ARMADILLO,

*Dasypus novemcinctus Texanus*

The average adult armadillo (see frontispiece)[3] is an animal with a body length of about eighteen inches, with a long, sharp nose, mulish ears, and a pointed tapering tail nearly as long as the head and body together. The most striking structural feature is the armor, which consists of a carapace composed of a solid, scapular shield anteriorly, a pelvic shield posteriorly, and a median banded region, consisting of nine movable bands of armor. There is a cephalic shield on top of the head and the tail is composed of rings of armor plate separated by armorless rings of soft skin. The

[1] M. Fernandez, *Morpholog. Jahrb.*, Bd. 39 (1909).

[2] H. H. Newman and J. T. Patterson, *Biological Bulletin*, XVII, No. 3 (1909).

[3] This illustration, painted by Mr. Kenji Toda from photographs of living animals, is the only really good figure of an adult *Dasypus* of any species that has been published. The usual figures are from photographs of stuffed specimens in entirely incorrect attitudes and in unnatural surroundings. This picture is in every respect an adequate representation of this interesting animal.

head with its long ears looks a little like that of a mule.
The feet are armed with heavy claws adapted for burrowing, and the legs, which are incompletely covered with
scales and hair, are comparatively short.  Many thou-
sands of the adult animals are slaughtered annually for
their armatures, which are shaped into baskets, the tail
being arched over and tied to the snout for a basket
handle.  Armadillo baskets are now fairly familiar
curios the world over.  By co-operation with the basket
merchants it has been possible to obtain an abundance
of material for embryological study without in any way
augmenting the slaughter of this species, which is going
on at an alarming rate.

The armadillo spends its life on the defensive and its
equipment for defense consists of numerous structural
and functional adaptations to a very special environment.  The armor is significant chiefly as a protection
from the thorns and spines of the arid vegetation in
which the animal feeds and into which it retreats for
shelter from enemies and from the tropic sun.  Doubtless, too, the armor is of service as a defense against
predacious enemies, but, if one may judge by the
armadillo's method of defense against hunting dogs,
the armor is less effective than the claws.  Stories of
the armadillo rolling up into a ball[1] when attacked
are totally inapplicable to this species, for the animal
turns over on its back and kicks viciously and effectively
with its powerful and heavily armed feet.  Although
an advocate of defensive armament, the armadillo
believes in active as well as passive defense.

[1] The little armadillo *Tolypeutes* is the only one that rolls up into
a ball.

Armadillos are pre-eminently insectivorous, although not exclusively so. Insect-hunting is carried on at night or at dawn and dusk. On warm nights one may hear the sniffing and grunting noises made by the armadillos as they root about among the dry leaves and ground vegetation after the manner of hogs. In the daytime they retire to their burrows, which are dug to a depth of six or seven feet in the dry soil. An enlarged chamber at the bottom is filled with dry leaves into which the animal burrows for warmth in the winter and during the cool spells of autumn and spring. It is in the burrow that the young are born and reared.

Mating occurs in October and the period of gestation is between four and five months. The young are quite advanced at birth and are able to walk about within the first few hours.

This incomplete account[1] of the natural history of the armadillo is given here merely to enable the reader to gain a slight acquaintance with the species that has furnished the embryonic material forming the central subject-matter of the present volume. In this place it may not be inappropriate to extend to the modest and retiring armadillo an acknowledgment of our indebtedness for much valuable data that no other animal could have supplied.

Our studies of the development of the armadillo cover the whole range of stages from ovogenesis to birth, with but one gap which, it is hoped, the near future will see filled in. It has been impossible so far

[1] A somewhat more adequate account of the natural history of this species may be found in the following paper: H. H. Newman, *American Naturalist*, XLVII (1913).

to secure the earliest cleavage stages of the normally developing egg, but a study of parthenogenetic cleavage,[1] as it occurs in atretic follicles, has been made; these data undoubtedly foreshadow some of the most funda mental events of normal cleavage. Until a study of normal cleavage is forthcoming the facts of partheno-genetic cleavage may be accepted as a temporary substitute.

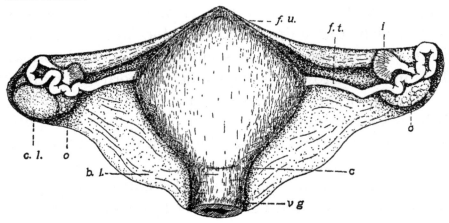

FIG. 3.—Uterus, ovaries, etc., of adult *Dasypus novemcinctus* (armadillo), showing simple squarish uterus with sharp fundus end (*f u*), cervix (*c*), Fallopian tube (*f t*), ovaries (*o*), only one of which, the left, has a corpus luteum (*c l*). (From Newman and Patterson.)

### THE FEMALE GENITALIA

The uterus of the armadillo is simple and quite like that of man and of other primates which produce but one offspring at a birth. Viewed from the dorsal aspect, the non-pregnant uterus appears to be broadly kite-shaped (Fig. 3), with the posterior angle blending into the vagina and with the right and left corners connecting with the Fallopian tubes. The free or fundus

[1] H. H. Newman, *Biological Bulletin*, XXV, No. 1 (1913).

end (*if u*) of the uterus runs to a point.   Internally two
grooves in the mucosa intersect at the very apex of the

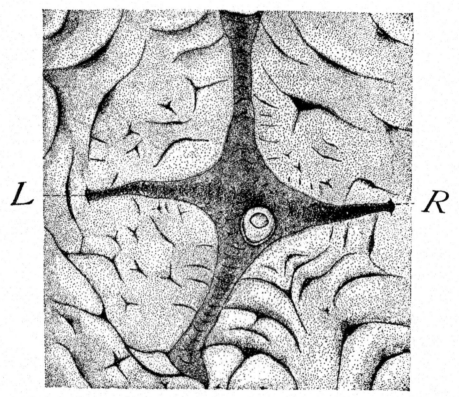

Fig. 4.—The small spherical body near the center of the cross-
shaped area is an armadillo egg beginning to adhere to the uterine
membranes in the fundus of the uterus.  The wrinkled, lighter area
surrounding the specialized attachment area is the general uterine
mucous membrane.  The lateral arms of the cross-shaped area are
grooves communicating with the right and left oviducts or Fallopian
tubes.  (From Patterson.)

fundus, and it is at this point, the center[1] of a cross-
shaped smooth area, that the egg attaches itself
when the placentation occurs (Fig. 4).  The whole

[1] If the egg has come from the right ovary, it will become attached,
as in the figure, somewhat to the right of center; if from the left ovary,
to the left of center.

mechanism is adapted to accommodate a single developing egg.

There is nothing suggestive of polyembryony about
the ovaries or oviducts. Each ovary, when no corpus
luteum is present, is kidney-shaped and about the
size of a small bean; in virgin females the two are of
the same size. In every pregnant female, however,
one of the ovaries is several times as large as the other,
owing to the presence of an enormous corpus luteum
in that ovary which has produced the fertilized ovum.
As in many mammals, the corpus luteum of pregnancy
is an extremely conspicuous object, especially in its
earlier phases, and its presence is unmistakable. One
of the earliest and most important discoveries in the
armadillo investigation was that *there is never more
than one true corpus luteum in the ovaries of a pregnant
female.* That the number of corpora lutea is a safe
criterion of the number of eggs involved in a pregnancy
is generally recognized by embryologists, and we may
feel safe in applying this test in cases such as those
offered by human and by ungulate twins where the
early embryonic history is unknown. The evidence
of the corpus luteum that, in the armadillo, only one
egg is produced and fertilized at a pregnancy is supported
by a study of ovogenesis, maturation, and fertilization.

### OVOGENESIS

The process of ovogenesis[1] (development of eggs)
in the armadillo presents in the earlier stages at least
nothing unusual, but is, on the contrary, quite typical
for mammals in general. Each of the young ovocytes

---

[1] H. H. Newman, *Biological Bulletin*, XXIII (1912).

(immature eggs) develops its own separate follicle, and the development of both ovocyte and follicle is much like that of the mouse or the cat. In only a very few instances has a follicle with two or more ovocytes been observed, and many ovaries totally lack double or multiple follicles. The full-grown ovocyte, which has a diameter of about 12 micra, is a little smaller than that of the cat and a little larger than that of man or those of rodents. Prior to maturation the definitive ovocyte of the first order lies in the discus pro-ligerus, a mass of follicular cells, which adheres to one side of the large follicular cavity.

A cytological examination of the full-grown ovocyte (Fig. 5) shows that there exists a pronounced cellular polarity.

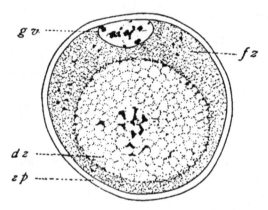

FIG. 5.—Full-grown ovocyte (un-maturated egg) of the armadillo, showing the mass of thin yolk in the center (*d z*) and the peripheral formative zone of protoplasm (*f z*), the nucleus or germinal vesicle (*g v*) at the animal pole, and the zona pelucida or egg shell (*z p*).

The germinal vesicle is flattened against the zona pellucida presumably at the animal pole. A comparatively homogeneous zone of darkly staining protoplasm which is thicker at the pole occupied by the germinal vesicle surrounds a sphere of coarsely vacuolated material, which must be identified as the deutoplasmic zone or yolk mass, in which are scattered bits of yolk material suspended in a fluid medium.

If the site of the germinal vesicle is that of the animal pole, the vegetative pole is that which is most nearly in contact with the yolk mass.[1]

During maturation an extremely radical change in polarity and general organization takes place. The comparatively homogeneous formative zone of protoplasm moves to one pole of the egg and forms a cap with a crescentic cross-section, thick in the middle and thinned out at the edges (Fig. 6). The originally central yolk mass moves to the pole of the egg opposite to that occupied by the formative cap, and comes to lie in contact with the zona pellucida for about one-third of the periphery of the egg. The germinal vesicle during this reorganization process enters the stage of a first polar spindle, which, peculiarly enough, lies tangentially to the periphery of the ovocyte and very close to the boundary between the formative and deu-

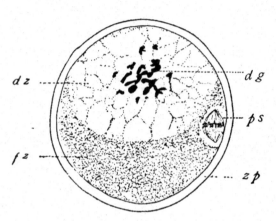

Fig. 6.—Maturating egg of the armadillo, showing total rearrangement of materials. The yolk (*d z*) with yolk granules (*d g*) occupies the animal pole, and the formative protoplasm (*f z*) occupies the opposite pole in the form of a cap. The nucleus (*p s*) is dividing to form the first polar body and lies tangentially to the periphery.

[1] This central position of the yolk and the peripheral position of the formative protoplasm are strikingly like those described by Hill (1910) for the marsupial *Dasyurus* and may therefore be interpreted as a primitive character.

toplasmic zones. This pronounced shifting about of materials within the maturing ovocyte must take place in a very brief space of time, for it has not been possible to find in a very large number of specimens examined any stages transitional between that before the reorganization begins (Fig. 5) and that after its completion (Fig. 6).

In attempting to interpret the significance of this remarkable cellular cataclysm it is interesting to note that Hill[1] finds exactly the same situation in the marsupial *Dasyurus*, which he interprets as a complete reversal of polarity. According to this writer the formative protoplasm now occupies the vegetative pole while the yolk mass occupies the animal pole. The position of the maturation spindle is in the formative zone, but as near the animal pole as possible. Whatever may be the morphological interpretation of the changes in cellular organization that take place during maturation, there can be little doubt as to the physiological significance of the shifts of material. If we may judge by analogy with *Dasyurus* and by a study of parthenogenetic cleavage in atretic follicles of the armadillo, the shift of the deutoplasm from the center to the periphery of the ovocyte is the first step in the process of deutoplasmic extrusion. This mass of thin degenerate yolk is of no value to the egg and must be voided before cleavage can begin. The process of voidance appears to be one involving rupture of the vitelline membrane and abstriction of the yolk, followed by a subsequent rounding up of the formative materials to form an egg that is much smaller than the original ovocyte and is presumably rejuvenated by the loss of its deutoplasm.

[1] J. P. Hill, *Quarterly Journal of Microscopical Science*, LVI (1910).

Prior to the complete abstriction of the yolk mass the egg completes its nuclear maturation. Studies of the chromosomes show that there are thirty-two in the ovocyte of the first order, and that after typical tetrad formation the number is reduced to sixteen in the first polar body and to sixteen in the ovocyte of the second order. The second maturation division is equational and produces a second polar body. It is very rare, however, to find a second polar body formed before the process of ovulation. This process probably occurs while the egg is in the oviduct just before fertilization. After a long search for tubal ova, I was fortunate enough to find one fertilized egg, apparently quite normal in every respect, in a part of the oviduct not far from the fimbriated funnel. This egg showed just two polar bodies and the male and female pronuclei lying close together in the formative protoplasm (Fig. 7). The deutoplasm had not yet been extruded. From this we might infer that yolk elimination occurs as an accompaniment of the first cleavage division, as in *Dasyurus*. A study of parthenogenetic cleavage[1] shows numerous stages in which the small yolkless egg lies within the zona pellucida surrounded by fragments of yolk. Later stages show very pretty cleavage spindles in the egg; clean-cut four-cell and somewhat doubtful eight-cell stages have been found. This is apparently as far as parthenogenetic development goes, so we must await the discovery of the events in normal cleavage before we can know with certainty what form of cleavage we have in the armadillo: whether it is regular, as in *Dasyurus*, or indeterminate, as in the

---

[1] H. H. Newman, *Biological Bulletin*, XXV (1913).

majority of eutherian mammals.   I am strongly inclined
to predict that when the cleavage is made known it will
prove to be much like that of *Dasyurus*.   This prediction
rests on two circumstances: first, that the history of
the ova of *Dasyurus* and that of *Dasypus* are alike as
far as we can trace them, and that the latter is unlike

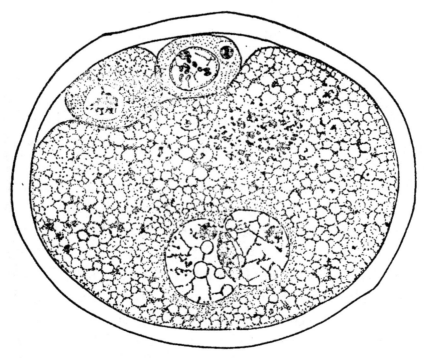

Fig. 7.—A fertilized armadillo egg with two polar bodies and the
male and female pronuclei side by side.

that of other species of *Eutheria;*   second, that the
arrangements of embryos in pairs and the mirror-
imaging effects that are so striking a feature of the
quadruplets are much more in accord with a type of
cleavage in which the blastomeres retain a regular and
definite position than with one in which the cleavage
cells shift about.

In general, we may say in concluding this account of the history of the egg up to the beginning of cleavage that there is nothing in any way suggestive of poly-embryony about any phase of the process. The eggs are ovulated singly, have small polar bodies, are ferti-lized by but one sperm, and begin cleavage in normal fashion, if we may judge by the data derived from a study of parthenogenetic cleavage.

### EARLY EMBRYONIC DEVELOPMENT

Fernandez[1] was the first to describe for any species of armadillo embryos in which no beginnings of poly-embryonic fission had taken place. He studied several early embryos of *Dasypus hybridus*, and although his material was in rather poor condition, it was clear that the blastodermic vesicle was a single individual in the stage examined and not unlike similar stages in some of the rodents. In the autumn of 1909 Newman and Patterson[2] obtained a number of stages of *Dasypus novemcinctus* covering the period from the primitive streak on to birth. In the following season a number of earlier stages were collected, some of the youngest of which were lost in fixation. At the end of this season these two collaborators divided forces on account of the removal of the senior author from Texas to Chicago. In the resultant division of the work the completion of the study of embryonic development was assigned to Patterson and the genetic problems to Newman. It took Patterson two more seasons to

[1] Fernandez, *loc. cit.*

[2] H. H. Newman and J. T. Patterson, *Journal of Morphology,* XXI (1910).

obtain the late cleavage and early embryonic stages, and to his paper[1] of 1913 we are indebted for a large part of the following account of this period of development.

The earliest stages found were collected on October 15; these consisted of two eggs in the Fallopian tubes, and several eggs floating freely in the uterine cavity. In no case was more than one egg found in the uterus or tubes of one female. From the fact that the Fallopian-tube eggs and all those found free in the uterus were in almost precisely the same embryonic stage, and from the additional fact that nearly every large female examined as late as three weeks after the earliest date mentioned had an egg in practically the same stage of development, it must be concluded that there is a *period of quiescence* of about three weeks, during which the egg either remains at a standstill or else develops so slowly as to make no perceptible progress. Although Patterson draws no conclusion as to the effects on development of this period of quiescence, it is my belief that this period holds the clue to the physiological explanation of polyembryony. Consideration of this point is deferred, however, until the general discussion of the underlying causes of twinning.

The following account of the embryology of *Dasypus novemcinctus* I have built up for the convenience of the reader around a number of carefully constructed figures; although slightly diagrammatic, they are accurate in all essential points, and are certainly more intelligible than microphotographs, or exact copies of preserved and sectioned material. I have found that zoölogists

[1] J. T. Patterson, *Journal of Morphology*, XXIV (1913).

in general have but little understood the essential features of polyembryonic development in this species. In the hope that this interesting piece of embryology may be made quite definitely intelligible to a general biological audience, I have had it re-illustrated by Mr. Toda. The first six figures (stages I–VI) are adapted from Patterson's photographs and figures; the remaining stages represent material to which I have had personal access.

*Stage I. The earliest embryos* (Fig. 8).—The youngest egg that has been found is in a rather late cleavage stage, in which the embryonic cells, eleven in number, form a small knot or inner-cell mass (*icm*) attached to the inner surface of the large, hollow sphere of non-embryonic cells, the trophoblast (*tr*). The trophoblast, as the name implies, has a purely nutritive function and serves later to attach the vesicle to the walls of the uterus. Of the eleven cells seen in the earliest egg, six differ from the others in having larger nuclei. These larger cells are destined to form the embryonic ectoderm. The other cells form the endoderm. Such an egg is in no way essentially different from any eutherian (higher mammalian) egg, and is unquestionably at this time without any visible indications of a prospective division into four embryos.

*Stage II. The beginnings of gastrulation* (Fig. 9).— The cells of the inner-cell mass have multiplied and have spread out into a flat disk of one or two layers in thickness. The distinction between ectoderm (*ec*) and endoderm (*en*) is now evident in that the less numerous endoderm cells are more deeply stained than the ectoderm cells. The endoderm cells are also beginning

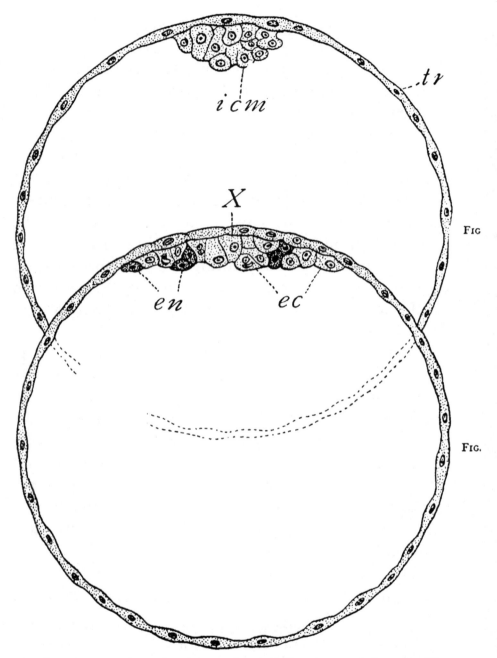

FIGS. 8, 9.—Sectional view of two earliest embryos of the armadillo. The two are shown overlapping, to save space. (For description see text, stages I and II.) The trophoblast (*tr*) and inner-cell mass (*ic m*), ectoderm (*ec*), and endoderm (*en*) are shown. The point *X* is the apical pole of the egg. (Modified from Patterson.)

to leave the part of the embryonic disk that is in contact with the trophoblast and to migrate downward to a position beneath the ectodermal mass.

*Stage III. Gastrulation completed* (Fig. 10).—The endoderm cells have now migrated away from the trophoderm and form a complete layer of somewhat flattened, deeply staining cells that lie in close contact with the compact mass of ectoderm cells (*ec*) on the side away from the trophoblast.[1] Very little change occurs in the trophoblast during the first three stages. Eggs of about the stage shown in Fig. 8 are found lightly attached to the uterine wall near the middle of the cross-shaped area shown in Fig. 4.

*Stage IV. Embryonic germ-layer inversion* (Fig. 11).—This stage is one of the most significant in the entire history in that it shows a curious inversion[2] of the normal relations of ectoderm and endoderm. We expect ectoderm to be outside and endoderm to be inside, but in the armadillo a sort of inversion occurs that results in the ectoderm getting inside the endoderm. Although this process is not necessarily followed by twinning, it at least appears to offer a highly favorable opportunity for this type of embryonic doubling.

Soon after the completion of gastrulation the somewhat flattened mass of ectoderm cells begins to round

[1] This method of gastrulation is very strikingly like that described by Hill for the marsupial cat *Dasyurus*, which is of interest when we recall that the remarkable changes in the ovocyte in this species are also like those of our armadillo. One may be fairly certain that the early cleavage stages of the two species will prove to be similar.

[2] A very similar type of germ-layer inversion has been described for several species of rodent. The work of Mellisinos on the mouse is especially interesting in this connection (*Arch. mikr. Anat. und Entw.*, Bd. 70) (1907).

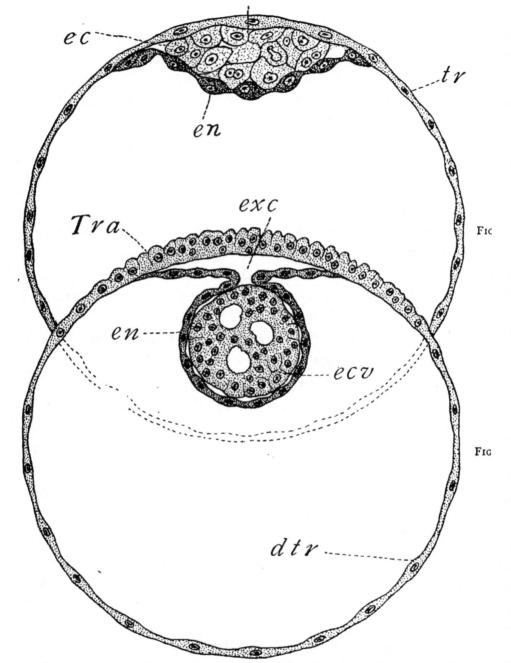

FIG

FIG

FIGS. 10, 11.—Two eggs of the armadillo, drawn as before, partly overlapping. The upper egg (stage III) shows the ectoderm (*ec*) rounding up into a ball and the endoderm (*en*) in the form of a continuous layer beneath the ectoderm. The apical pole is at *X*. The lower egg (stage IV) shows the ectoderm (*ec*) rolled into a ball and nearly surrounded with endoderm (*en*). Trophoblast has been thickened into Träger (*Tra*) at the upper pole and the remainder is called diplotrophoblast (*dtr*); extra-embryonic cavity (*ex c*). (Modified from Patterson.)

up and the process of rounding up is unquestionably
a sort of invagination of the middle or apical portions, so
that the free edges unite above and the whole mass
becomes essentially a hollow ball.  In Fig. 11 the ball
appears to be only partially hollow, but the ectodermic
mass is morphologically a vesicle with a central cavity,
which is the primitive amniotic cavity.  This invagination of the ectoderm involves a very fundamental
reversal of the primary axis of the embryonic
materials, for the head end[1] of the ectodermic mass is
now directed away from the original animal pole of
the egg.

The ectoderm is the active agent in this process of
germ-layer inversion, and the endoderm plays the
merely passive rôle of maintaining its contact with
the ectoderm.  The result is that it comes almost
completely to surround the ectodermic vesicle.  As a
sequel to the inversion process the ectoderm becomes
totally separated from contact with the trophoblast,
and a cavity arises between the latter and the embryonic
tissues, which is the beginning of the extra-embryonc
cavity (ex c).

That part of the trophoblast which has adhered
to the uterine mucosa has at this period entered upon
a process of rapid cell proliferation preparatory to
invading the maternal tissues.  At this time only short
protrusions have been formed that serve as mechanical
aids to adhesion.  This specialized region of the trophoblast is destined to form the primitive placenta or

---

[1] It may be noted here that in later stages (Figs. 15, 16, and 17) the
heads of all embryos are directed away from the original apical end
or animal pole of the egg, which is the attached end.

Träger, while the thin-walled part of the trophoblast is known as the diplotrophoblast (*dtr*).

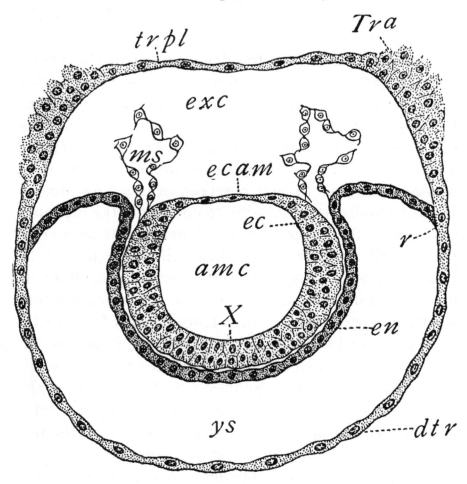

FIG. 12.—An armadillo egg (stage V) attached by the Träger to uterus, and shown as if torn away at *Tra*. The ectoderm (*ec*) is a hollow vesicle. The apical pole is at *X*. Endoderm (*en*) joins diplotrophoblast (*dtr*) in a ring (*r*). The trophoderm plate (*tr pl*) lies within the Träger collar (*Tra*); ectodermal layer of amnion (*ec am*); amniotic cavity (*am c*); mesoderm (*ms*); much enlarged extra-embryonic cavity (*ex c*). (Modified from Patterson.)

*Stage V. The period of rapid growth and the establishment of bilaterality* (Fig. 12).—During the earlier stages very little increase in the actual mass of tissue has taken place. The egg has merely increased in diameter owing to the accumulation of fluid in the trophoblast cavity.

Simultaneously with the development of the Träger and its invasion of the uterine mucosa there begins a period of rapid cellular proliferation and consequent tissue growth, evidences of which have already been noted in Fig. 11. At the stage shown in Fig. 12 the Träger has deeply invaded the maternal mucosa by a process analogous to that observed in invading tumors. It has been deemed advisable to omit from the drawings that part of the Träger that has penetrated the maternal tissues and to show by broken cellular contours the points (*tr*) where the vesicle has been severed from its nutritive connection with the mother. A vesicle like that shown in Fig. 12 is more than twice as great in diameter as that shown in Fig. 11, and the increase in size is due in part to the marked enlargement of the extra-embryonic cavity, in part to the expansion of the cavity of the ectodermic vesicle, which is now a true amniotic cavity (*am c*). The subspherical ectodermic vesicle has thinned out on the side toward the extra-embryonic cavity to form the ectodermal component of the amnion (*am*). The embryonic ectoderm is now a vesicular mass of cells somewhat elongated in the bilateral axis of the uterus and with anterior or apical end at *X*. The embryo is really a gastrula turned inside out, and hence the process deserves the name "germ layer inversion." If the amnion were cut and the germ-layers were reinverted, we should get a normal gastrula with the apical point up and the basal parts down. It should be emphasized that the embryo though inside out is clearly *polarized* and *bilateral* and that it is still *one* embryo.   A further evidence of bilaterality is seen in the mesoderm (*ms*)

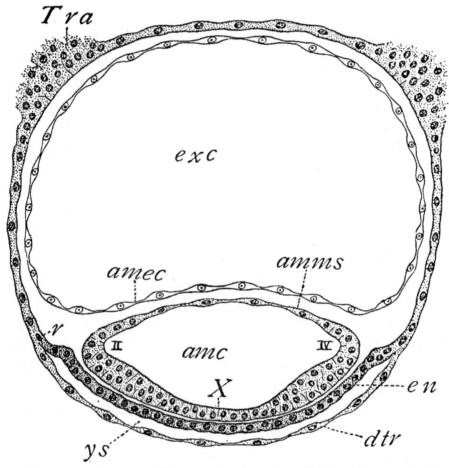

FIG. 13.—An armadillo egg showing first division into a double individual.   The two first embryos (II and IV) are shown as right and left outgrowths of the ectodermic vesicle.   (For details see description under stage VI.   Lettering as in Fig. 12.)   (Modified from Patterson.)

which is proliferating at two bilateral points where the ectoderm and the endoderm part company.

*Stage VI.   The first step in twinning—the primary embryos* (Fig. 13).—The ectodermic vesicle, which was originally situated at or near the animal pole of the egg, has progressively retreated from this pole and now lies with its apical part (head end) almost at the opposite (vegetative) pole. The formerly voluminous cavity, shown in stages I–IV, which we have called the trophoblast cavity, has gradually diminished in relative and absolute volume (stages V and VI) until it is merely a flat crevice (ys) between the endoderm and the trophoblast. At this stage the free edges of the endoderm have fused with the trophoblast at r, and that portion of the trophoblast distal to the ring of fusion (the diplotrophoblast) has thinned out preparatory to a subsequent total disappearance, as in stage VII. The retreat of the ectodermic vesicle from pole to pole and the crowding out of the original trophoblast cavity appear to be due to the pressure of the rapidly enlarging extra-embryonic cavity (ex c), which is now lined internally with a complete vesicle of mesoderm (ms). That part of the mesoderm next to the ectodermal amnion completes the amnion proper. No embryonic mesoderm is as yet formed.

The ectodermic vesicle is seen to be flattened against the endoderm, and two hollow evaginations are shown at right and left sides; these are the primordia of the *primary embryos* (II and IV). These outgrowths constitute twin embryonic areas with the apex or head end of each pointing toward the apex of the ectodermic vesicle (X), and with the posterior or growing end of each pointing the one toward the right and the other toward the left side of the uterus. It seems quite evident that the bilaterality of this twin embryonic vesicle has been secondarily imposed upon it by the bilaterality of the uterus, for we can see no other way of explaining the coincidence that exists between the uterine and embryonic axes. No important change has occurred in the Träger ring except that further invasion of the uterine mucosa has continued. In some eggs the disk of the trophoblast (*tr d*) lying within the Träger ring appears to be quite free from the uterine mucosa and to form the boundary of a more or less extensive fluid-filled cavity, which had been called by Fernandez the Träger cavity. This cavity is evidently of little morphological significance and may be ignored in subsequent stages.

*Stage VII. The origin of quadruplets. Secondary embryos formed* (Fig. 14).—It is at the stage shown in Fig. 14 that the second step in twinning occurs, but the figure, because it is a bilateral sectional view of the egg, fails to show the *secondary embryos*. The primary embryos II and IV lie respectively to the right and to the left of the egg, while a shorter secondary embryonic outgrowth appears to the left side of each primary embryo, so that the two secondary embryos lie with their axes pointed one toward the dorsal and the other toward the ventral side of the uterus. The embryo on the dorsal side is called III and is said to be the secondary embryo paired with the primary embryo IV, while the ventral secondary embryo is called I and is similarly related to the primary embryo II. The outline sketch (Fig. 15) shows the axes of the four embryos, seen from the distal end of a vesicle like that shown in Fig. 14. Note that there are evidences of tertiary outgrowths between I and IV and between II and III. In *Dasypus hybridus* such outgrowths evidently form embryos, for

the typical number of embryos in that species ranges
from seven to twelve. An inspection of Fig. 15 leads
one to a different interpretation of the relation of

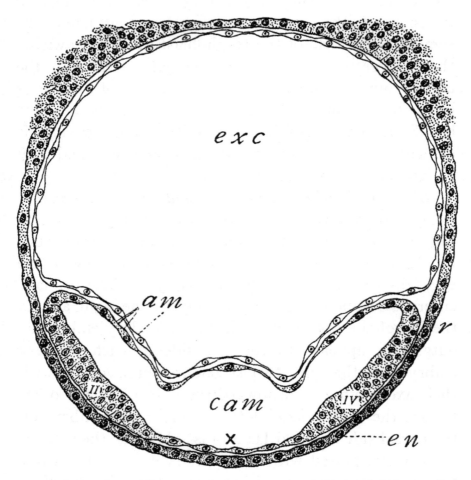

FIG. 14.—Armadillo egg showing two out of four embryos growing
away from the common amnion (*c am*). The other two embryos
(I and III) do not show in this plate. A view from the lower pole of
the egg is seen in Fig. 15. (For description see stage VII.)

secondary to primary embryonic primordia from that
offered by Patterson. I am inclined to believe that
prior to the visible formation of outgrowths the ecto

dermic vesicle had already been physiologically differentiated into a number of radially arranged equivalent
apical points focusing toward the original common apex,
and that those particular apical points which happened
to be directed respectively to the right and left sides of
the uterus had more room in which to grow and consequently developed more rapidly than those located in
other sectors, and thus
became the primary
embryos. The less
favorably situated
points that grow less
rapidly are secondary
and tertiary in time,
but genetically they are
as independent and as
old as the primary
embryos. This expla-
nation of the curious
bilateral orientation of
the quadruple vesicle
in the uterus accords
with the facts of devel-
opment and of heredity
better than others

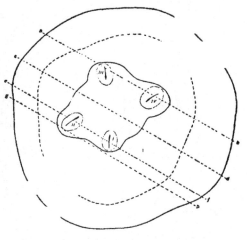

Fig. 15.—Outline of lower pole of
stage like Fig. 14, showing the four
embryos. The dotted lines across
are lines of certain sections not shown
here. The four embryonic areas are
numbered I, II, III, and IV. (From
Patterson.)

which have been previously offered, and removes, it
seems to me, much of the mystery involved in the
problems arising from a study of resemblances and of
symmetry relations among the quadruplets.

*Stage VIII. The retreat of the embryos from the
common amnion toward the original animal pole of the
egg (Fig. 16).*—In order to bridge over the gap between

the stages shown in Figs. 15 and 17, the diagrammatic Fig. 16 is shown. Here the embryos, which are in an early primitive-streak stage, have migrated down the meridians of the vesicle and have left behind them thin-walled amniotic connecting canals that still anchor the head ends of the embryos to the small common amnion (*c am*) situated at the distal or vegetative pole of the egg. The posterior end of each embryo is now growing rapidly toward the rim of the Träger ring, a junction with which is soon to be established.

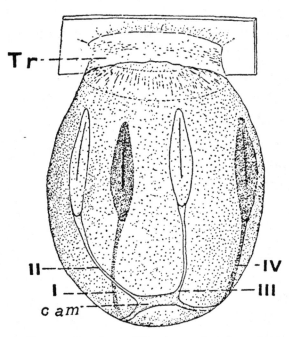

FIG. 16.—Armadillo egg showing four embryos in early primitive streak stages. (See stage VIII.) (From Patterson, but inverted in order to be comparable with the other stages.)

The sectional view shown in Fig. 14 indicates that the heads of embryos II and IV are growing apart, leaving between them a thin sheet of ectoderm. This, together with the median portion of the original amnion, is destined to form a vesicle that remains connected by canals with the amnia of the individual embryos and is therefore called the common amnion (*c am*).

That region of the original trophoblast which is called diplotrophoblast and lies at the distal pole of the vesicle and within the ring formed by the fusion of the endoderm and the trophoblast proper has now entirely disappeared and the distal portion of the vesicle is bounded externally by endoderm. The point of fusion between the endodermic and trophoblastic parts of the vesicle wall is not easy to detect, but the endoderm is likely to be more deeply stained in microscopic preparations.

As the embryonic ectoderm grows down the sides of the egg toward the Träger it carries with it the adjacent endoderm, so that subsequently a large proportion of the vesicle comes to be covered with endoderm.[1]

*Stage IX. The attachment of the quadruplet embryos to the Träger* (Fig. 17).—The figure is redrawn from one published in 1910.[2] The aim has been to represent the vesicle as a transparent object, and this has been, partially at least, realized. In the previously published figure the axis of the vesicle was incorrectly placed so that the heads of the embryos were directed toward the top of the page. In all previous figures the original animal pole of the egg is toward the top of the page, and in order to preserve this arrangement with consistency, all figures must show the Träger end of the egg at the

[1] Reference to Fig. 27, which is taken from one of Fernandez' figures illustrating conditions in the *Mulita* (*D. hybridus*), will serve to show the extent of the peripheral endoderm. It also makes clear the relations of the germ-layer components of the entire vesicle. This figure serves equally well for our species, *D. novemcinctus*.

[2] H. H. Newman and J. T. Patterson, "Development of the Nine-banded Armadillo from the Primitive Streak Stage to Birth," *Journal of Morphology*, V, 21.

top and the common amnion at the bottom of the figure. This arrangement, doubtless, looks upside down to one familiar with previous accounts of the embryology of

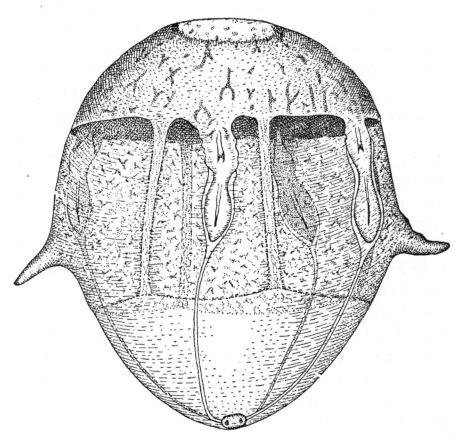

FIG. 17.—Armadillo egg with quadruplet embryos attached to primitive placenta, which is a bowl-shaped area at top of figure. Note the connecting canals running downward to the original common point of origin, which is now occupied by the small common amnion. (See stage IX.) (Redrawn from Newman and Patterson.)

*Dasypus*, but the reversal of axis is an important feature of the embryonic history and must be indicated just as it appears. The comparatively smooth outline of the Träger region is due to the fact that the period of

Träger nutrition has passed, the Träger rim has disappeared, and the egg is beginning to develop ridges which act as physiological drill points and enable the young villi to penetrate the maternal mucosa. A small circular area at the very top of the egg is comparatively free of villous ridges and is destined to become a large non-villous area such as is shown in Fig. 19.

Each embryo is in a somewhat advanced primitive-streak stage and has formed a broad, bandlike connection with the margin of the developing placenta, a connection that is the forerunner of the umbilicus. A short endodermal allantois opens externally and extends inward toward the placental margin, though it is not destined to play any important rôle in placentation; it is evidently a vestigial structure. The relation of this structure to the umbilicus is better shown in the sectional figure of a little later stage (Fig. 26). Each embryo occupies its own extra-embryonic area and is isolated from corresponding areas of adjacent embryos by a sort of partition resembling the sinus terminalis of an avian embryo. The extra-embryonic areas are covered with blood islands which are the forerunners of the system of vitelline blood vessels seen in the next figure (Fig. 18).

Each embryo has its own amnion which is connected by its amniotic connecting canal with the small common amniotic vesicle, whose origin has been already described.

The two lateral horns seen in the figure are protuberances of the egg membranes that are forced into the openings of the paired Fallopian tubes. These points are very useful as a means of orienting the vesicle with reference to the uterine axes; by means of

these two points we are able to show that the arrangement of embryos is not precisely in accord with the
uterine bilaterality. Embryos II and IV are not
exactly lateral, nor are embryos I and III strictly ventral
and dorsal respectively. About a third of the egg at
the distal (lower) end is very thin-walled and transparent, owing to the fact that it is composed—except
where the amnia are fused with it—of but two layers of
cells, endoderm externally and mesothelium internally.
One can look into this part of the egg as through a
window.

*Stage X. Five- and seven-somite embryos in an egg
with early true placenta* (Fig. 18).—Several points of
interest are shown in this figure, which is redrawn from
one previously published in which an inadvertent error
was made.[1] The placenta now consists of a broad band
of villi that had penetrated the mucosa to such an
extent that it required some effort to pull the egg free.
The villi are rather flat and tend to overlap like shingles.
Each embryo is connected with the placenta by means
of a double primitive umbilicus, which as yet is not
traversed by blood vessels. The allantois is, as before,
vestigial. The two primary embryos (II and IV), those
nearer the lateral edges of the blastocyst, are more
advanced than the two secondary individuals (I and
III); this is due to the fact that the former are somewhat older than the latter. The bilateral arrangement
of the egg is still preserved, as in stage IX, by the
lateral horns or bumps that represent the points of
protrusion of the vesicle wall into the Fallopian tubes.

---

[1] In drawing in the details of the embryos the two upper embryos
were formerly reversed in position. This redrawing corrects the error.

Vitelline blood vessels, arranged somewhat as in the area pellucida and area opaqua of the avian egg, are

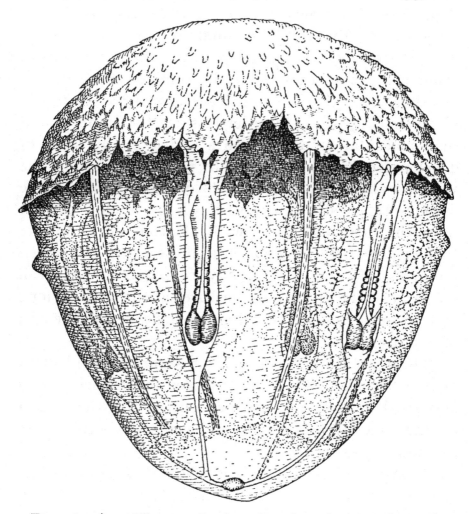

FIG. 18.—Armadillo egg showing that the primary embryos (II and IV) are in advance of the secondary embryos (I and III). The primary placenta as Träger is becoming displaced by the secondary true placenta, which is covered with villi, or short finger-like processes (see stage X). (Redrawn from Newman and Patterson.)

very characteristic features of this stage; no blood is found in this vitelline area and no vitelline circulation is ever established. The common amnion and the amniotic connecting canals still persist and serve by their forked arrangement to indicate the pairing of embryos, for the amniotic canals of the two embryos derived from one side usually unite into a short single canal just before they enter the common amnion.

*Stage XI. A middle-aged egg showing four discoid placentae* (Fig. 19).—In order to show the embryos and their placental connections in a vesicle of this stage of development, it is necessary to remove a part of the vesicle wall together with parts of the placenta. The drawing represents an egg with a large window cut out from the median ventral wall. Dotted lines indicate those parts of the two placental disks which have been removed; nothing else of any consequence has been removed. The entire egg measures about 35 mm. long and 30 mm. wide; each fetus has a head–rump length of about 14 mm.

The placenta consists of four separate ovoid disks of treelike villi. The disks of paired fetuses are closely appressed but visibly independent, while there is a distinct space both dorsally and ventrally between the placental disks.

A large area occupying the whole upper part of the egg has become essentially non-placental, although sparse villi are seen dotted over its surface, especially near the margins of the specialized placental areas.

The four fetuses are seen to be clearly paired, one pair facing toward the right and the other toward the left; there is much more space between the umbilical attachments of unpaired than between paired individuals.

The upper fetus on the left is fetus I, its partner is
II; the lower right fetus is III and its partner is IV

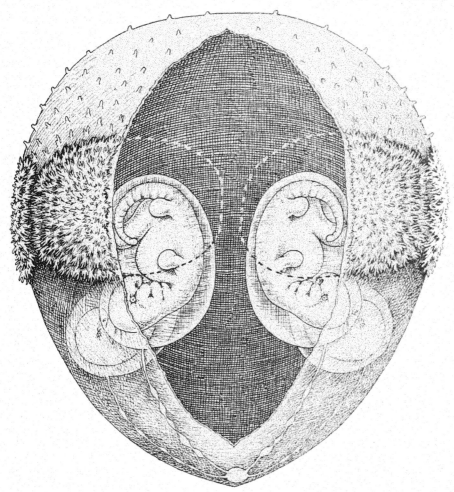

Fig. 19.—An armadillo egg about six weeks after fertilization
showing the two pairs of fetuses, revealed by the removal of part of the
egg membranes.   Each has its own oval placental area, its own amnion,
umbilicus, etc.   The heavy dotted lines indicate the boundaries of
the removed portion of the placental disks of the two nearest embryos.
(See stage XI.)

Each fetus has its own amnion, but at this time the
amnia occupy only a small part of the total volume

of the blastocyst cavity. The shrunken amniotic connecting canals and the common amnion are quite obvious at this time and persist in stages still more advanced than this. The large area at the bottom of the vesicle, extending from the placenta to the somewhat pointed vegetative pole, retains its transparency.

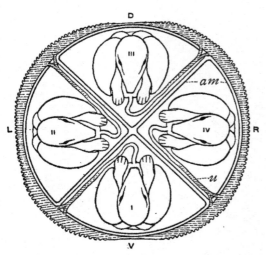

FIG. 20.—Semi-diagrammatic sectional view of full-term armadillo egg, seen from the lower pole (cf. Fig. 19). The four fetuses occupy four quadrants of the egg, with partitions between them composed of the fused amnia of adjacent individuals. The placental areas are two thick bilaterally arranged double disks, each with the umbilical cords of two fetuses attached. (See stage XII.)

It may be noted that from this stage up to a stage shortly before birth no changes occur that are especially significant for a study of twinning. The whole vesicle grows enormously, but the embryo and the amnia increase in volume more rapidly than does the vesicle wall. The result is that the amnia fill the cavity of the blastocyst as is shown in the next figure.

*Stage XII. A full-term quadruplet egg* (Fig. 20).— The figure of the vesicle is semi-diagrammatic in detail,[1]

[1] Strahle, in a paper entitled "Über den Bau der Placenta von *Dasyurus novemcinctus*," gives an incorrect account of the arrangement of the fetuses in the uterus, in that he shows the amniotic partitions coinciding exactly with the dorsoventral and the bilateral axes of the uterus.

but accurate in so far as the relation of placenta and membranes is concerned. The view is one that would be obtained if one looked into the vesicle from its lower transparent end; the eye sees the embryos head-on, so to speak. The axes (dorsoventral and bilateral) are the same as those of the mother.

In full-term eggs the placental complex consists of two well-defined areas corresponding to the right and left sides of the uterus. Those two heavily villous areas ($p'$ and $p''$) are separated, dorsally and ventrally, respectively, by narrow non-villous areas ($d\,n\,v$ and $v\,n\,v$); these serve as points of demarkation between the right and left placental regions; they are not always precisely mid-dorsal and mid-ventral, but usually slightly asymmetrical. Frequently placental blood vessels run across the non-villous area so as to connect the circulation of the two sides. A single fetus may have parts of its placental material in both of the bilateral placental areas. It appears then that the double placenta is merely a mechanical adjustment of the fetal membranes to the bilateral blood supply of the uterus.

Although one fetus appears to belong to the right quadrant, another to the left, a third to the dorsal, and a fourth to the ventral, it is obvious that, so far as placental connections are concerned, one pair (I and II) belongs to the left side, the other pair (III and IV) belongs to the right.

The umbilicus of each fetus is close to an amniotic partition composed of the right-hand side of its own amnion and the left-hand side of the amnion of its right-hand neighbor. Curiously enough there has been a crowding to the left on the part of each amnion until each one has filled a full quadrant of the vesicle and has fused with adjacent amnia wherever contact has been established. Why the pushing over of the amnia always goes toward the left (anti-clockwise) and never to the right (clockwise) is a problem in developmental mechanics for which I have no solution. Evidently, however, the vesicle is still acting as a unit and responding to the same predetermined bias toward the left which was expressed in the period of embryonic segregation when each primary individual always pairs with a secondary individual at its left side.[1]

An egg at full term has a transparent area at both proximal and distal ends and the broad but broken placental zone about the equatorial region. It is readily seen that, when the arrangement of fetuses is so diagrammatic, there can be no difficulty in preserving their paired arrangement. One has merely to open the vesicle at the point ($v\,n\,v$) in each case in order to be able at any subsequent time to identify fetuses I, II, III, and IV. This situation is, of course, highly favorable for the study of correlation and heredity, as will presently appear.

*Atypical numbers of fetuses in D. novemcinctus.*—The description of embryonic development herewith presented applies to a large proportion of cases. In about 3 per cent of cases, however, the diagrammatic relations typical for the species are distorted by the development of more or less than the typical number of embryos.

[1] It is only natural to refer in this connection to the asymmetry present in the maturating ovocyte (Fig. 4), where the maturation spindle is seen to occur at one side of the formative zone. There may be established here the basis of a growth bias toward the left.

Fernandez has described and figured a rare case in which a single fetus occupied an egg alone. The placental relations were decidedly irregular in that the villous area was an incomplete irregular zone or band partially encircling the equator of the vesicle. It seems evident that this condition is not due to a sporadic reversion to a uniparous condition but to the precocious death of several embryos. The zonary position of the placenta seems to require this interpretation, for, if the condition is due to a sporadic recurrence of a type of development typical for uniparous species of armadillo, the placenta should be discoid and at the proximal pole of the vesicle.

In my own collection of over two hundred eggs there occurred one case of twins which were born in captivity, and whose placental relations were not determined; two sets of triplets, in one of which unmistakable traces of a fourth embryonic rudiment appeared; and three sets of quintuplets, in one of which an additional sixth degenerating embryonic rudiment was evident. In all of these cases there was a pronounced lack of adjustment to the uterine axes.

One may therefore conclude that the *quadruplet* condition is practically specific for *D. novemcinctus*, and that there is little evidence that progressive evolution will either augment or decrease the typical number.

*Variations in the relations of embryos.*—In over 75 per cent of cases the paired arrangement of embryos and their rather precise method of union with the right and the left placental disks is like that described in connection with stage IX, Fig. 17. When I first studied the orientation of embryos with reference to the uterine bilaterality, I was much impressed with the regularity of arrangements; but, after a careful examination with the idea of finding out just how definitely fixed this kind of orientation is, I am surprised to discover a considerable range of variability in fetal relations.

One not infrequently finds the point of attachment of fetuses II or III much nearer the mid-dorsal line than that shown in Fig. 20; similarly, fetuses II or IV may be much nearer the mid-ventral line. In such cases the placental vessels of the fetus which is attached near the edge of one or other placental disk invade both right and left placentae. This may be readily shown by injections.

Cases of this sort point to the conclusion that the apparent bilaterality of the blastodermic vesicle is simply a semblance of bilaterality which has been imposed upon a more fundamental symmetry of the vesicle itself by the uterine blood supply. The two heavy placental disks that develop respectively on the right and left sides of the uterus occupy their positions because it is in those places that placental growth is favored by the proximity of the uterine bilateral blood trunks. The embryos are usually paired so that one pair draws nutriment from one placental disk and the other pair from the other disk; but there are many exceptions, as has already been indicated.

In previous papers I have laid much stress on the paired arrangement of embryos and have been inclined to underemphasize those cases in which pairing was absent; a few cases of non-pairing are accordingly cited. In some advanced sets it is noteworthy that, when the ventral bridge between the two lateral placental areas is severed, three fetuses appear to be attached

to one large lateral disk and one fetus to a smaller disk
on the opposite side.    Again, in two sets (Figs. 21 and
22) that are in approximately the stage shown in Fig. 19,
an interesting irregularity appears in the arrangement
of the amniotic connecting canals with reference to the
common amnion.    Instead of having the usual dicot-
omous branching on each side, there are in both cases
three connections on one side and a single unbranched

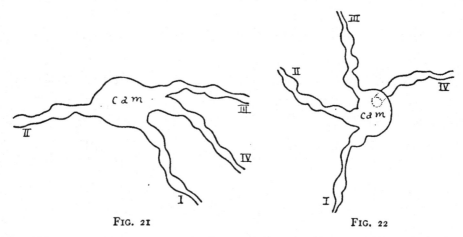

FIG. 21                              FIG. 22

FIGS. 21 and 22.—Outline views of the common amnion and the
connecting canals of fetuses.  These were drawn from two eggs that
were in stages between Figs. 18 and 19.  In Fig. 21 the canal of fetus II is
separated far from its partner, fetus I.  In Fig. 22 fetus IV seems to
have originated separately and does not seem to be paired with fetus III.
These are cases of non-pairing and may represent a not uncommon
condition.

connection on the other.    Doubtless if a larger collection
of equivalent stages were available other similar conditions would be revealed.    This departure from the
symmetrical paired arrangement is significant when
compared with what Fernandez describes for the
*Mulita*, where the irregular condition is the rule and
there is little evidence of pairing (compare Figs. 27–31).

One might have been led to suspect the occurrence of exceptions to the rule that the embryos of *D. novemcinctus* consist of two pairs, one derived from the right and one from the left primary ectodermic outgrowth, for, long before the information just given was available, it was known that the inter-resemblances of quadruplet sets were occasionally out of accord with the paired arrangement. Sometimes three out of four fetuses were alike in the possession of some one genetic feature and the fourth was different; similarly, it was occasionally noted that but one fetus possessed a peculiarity and the other three were without it. At the risk of anticipating the conclusions expressed in a subsequent chapter I may say that, were we in possession of the facts as to the origin of the four quadruplets from the common ectodermic vesicle, all of the data on resemblances, which are to receive treatment subsequently, would be completely rationalized.

### DEVELOPMENT OF THE PLACENTA

The history of the placenta is of considerable importance for an understanding of twinning. It gives us the criteria for distinguishing the four embryos and their paired or non-paired relationships even in the last stage of uterine development. A brief résumé of the facts relating to the placenta will accordingly insure a clearer understanding of what is to come. The primary placenta is the Träger, a ring of specialized trophoblast that forms the first connection with the uterine mucosa. This Träger subsequently, as in Fig. 17, becomes uniformly studded with adhesion pads destined to become the burrowing tips of the

definitive villi.   Later, when the four embryos form
an attachment at their posterior part with the Träger
ring, a co-operation between embryonic endoderm
and the Träger epithelium takes place, resulting in the
formation of hollow vascular villi that invade deeply
the maternal mucosa.   Four separate patches of villi
appear corresponding to the point where each embryo
has developed its own nutritive attachment with the
mother.   With the rapid peripheral extension of those
villous patches there is an apparent fusion of the four
placentae into a scalloped placental ring which appears
to be continuous, but is not, for there is no admixture
of blood between adjacent fetuses.   This has been
fully demonstrated by injections.   This "compound
zonary placenta," as it has been called, remains as an
apparently complete ring about the equator of the
vesicle until rather late stages of development.   Then
there follows a separation of the ring into two almost
separate discoid placentae, each of which receives the
umbilical cords of a pair of embryos.   If one examine
a gravid uterus during the last month of pregnancy, he
will readily note that in the median dorsal and median
ventral regions there is an area almost devoid of placental tissue.   If the uterus be cut open along this clear
line on the ventral side, the cut will fall between the
placental disks of the two halves of the uterus, and the
orientation of pairs will usually be preserved.

# CHAPTER III

## MODES OF TWINNING IN OTHER SPECIES OF ARMADILLO

Before attempting a discussion of the probable origin and causes of polyembryonic twinning in the nine-banded armadillo it will be well to learn something about the conditions known for other species of armadillo. Nothing is more likely to furnish a clue to the solution of the problems presented by one particular species than a comparative study of the embryology of allied forms.

Previous references have been made to Fernandez' account of the polyembryonic development of the *Mulita* (*Dasypus hybridus*). His account of stages corresponding to those illustrated by our Figs. 12–19 are so nearly identical with those described in the last chapter for *D. novemcinctus* that it will be unnecessary to do more than indicate the somewhat minor differences in detail. A comparison of Fernandez' figures (Figs. 23 and 24) with our Figs. 12 and 14 will indicate the close similarity.

The two species of *Dasypus* are evidently very closely related, and it would appear probable that *D. hybridus* is a comparatively recent derivative of *D. novemcinctus*. The only important particular in which a difference exists is in the number of young in a polyembryonic litter. While in *D. novemcinctus* the number is almost uniformly four, in *D. hybridus* the number is

68

larger though much less definitely fixed. There is a strong tendency, however, for the species to settle down upon the number eight, though litters of nine are frequent and from seven to twelve are reported.

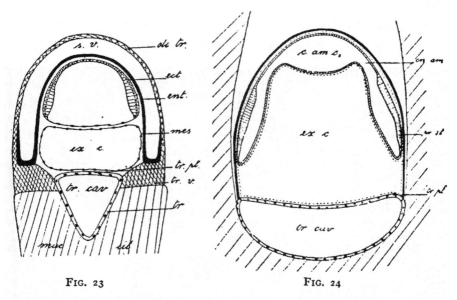

FIG. 23                    FIG. 24

FIGS. 23 and 24.—Diagrammatic views of two stages in the development of *Dasypus hybridus* (the *Mulita* armadillo). For comparison with equivalent stages of *D. novemcinctus* (Figs. 12 and 14) they should be viewed inverted. Note Träger cavity (*tr cav*), trophoderm plate (*tr pl*), ectoderm (*ec*), endoderm (*en*), mesoderm (*ms*), diplotrophoblast (*dtr*), extra-embryonic cavity (*ex c*), uterine mucosa (*muc ut*), common amnion (*c am*), primitive streak of embryos (*pr st*), amniotic connecting canal (*cn am*). (From Fernandez.)

The fact that there is so much variability in the number of polyembryonic offspring in this species apparently indicates that the condition is of comparatively recent origin. This view is supported by the fact that so large a percentage of embryos in advanced stages are dead or show signs of marked abnormality, owing to overcrowding and ill-success in the struggle for placental surface. What evidently happens is that four or more secondary growing points start to develop simultaneously, instead of the two that are characteristic of *D. novemcinctus,* and that normally each of these growing points divides into the primordia of two embryos; but sometimes more than two embryos are the result of this fission and sometimes

FIG. 25.—Photographic view of a set of embryos of *D. hybridus* (after Fernandez). Note the common amnion in the middle and the amniotic connecting canals running to the nine embryos.

no fission occurs. Such irregularities as these are similar to the formation in *D. novemcinctus* of three embryos instead of the typical two from one-half of the ectodermic vesicle, resulting in five embryos. That the above interpretation of the origin of the number of fetuses in *D. hybridus* is probably correct may be inferred from an examination of Fernandez' photographs; Fig. 25 is taken from one of these. Note that the amniotic canals are forked just as are those of a pair of embryos of *D. novemcinctus* that come from a primary outgrowth, a fact that lends probability to the view that the development of polyembryony in the two species of *Dasypus* is practically identical in character. It should also be said that, even in sets with eight or more fetuses, there is no exception to the rule that all from a single egg are of the same sex. Unfortunately nothing is known about the heredity of armor characters, nor about the symmetry relations existing between the different members of a polyembryonic set. Such a study of the species, if correlated with what has been published in these connections about *D. novemcinctus* would be important.

We are indebted to Fernandez for a clear understanding of the interrelations of germ-layers and of the peculiarities of amnia and allantois in *D. hybridus*. The diagram (Fig. 26) is adapted from Fernandez' first paper. It would serve, however, almost equally well for *D. novemcinctus*. Only two embryos that lie to right and left of the egg are shown. Note the external endoderm (*en*), the rudimentary allantois (*al*), and the short yolk stalk opening into the yolk sac, which is in this case inverted so as to form the external layer of the vesicle. The belly-stalk (*bs*) or primitive umbilicus is establishing a relation with the Träger, but as yet no umbilical vessels have invaded it. A posterior prolongation of the amnion (*p am*) goes back toward the Träger, but appears to have no part in the formation of an umbilicus. The amniotic connecting canals and common amnion are shown with both ectodermal and mesodermal layers present.

Five years after his first paper on polyembryony in
*D. hybridus* Fernandez[1] reported further facts about

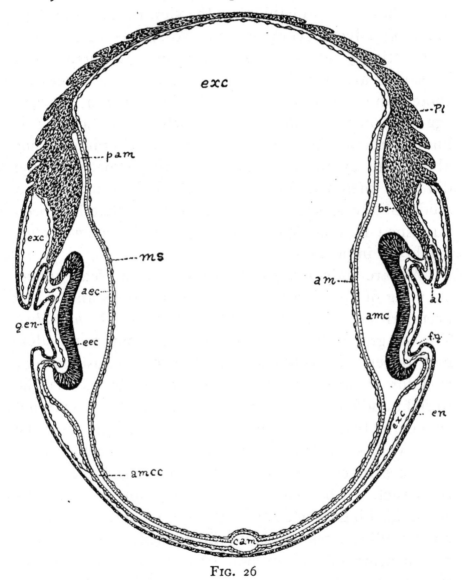

FIG. 26

this species and dwelt especially upon the origin of the
individual embryos from the undivided egg.  He appears

[1] M. Fernandez, *Proc. IX^e Congrès. de Zoöl. Monaco,* 1914.

to have adopted the "budding" hypothesis as an explanation of the mode of isolation of the several embryos
from the ectodermic vesicle, since he uses the word
"Sprossung" to describe the process. Possibly, however,
this term may merely imply evagination or outgrowth,
and thus be free from the unfortunate implication carried
by the word "budding." In brief, this is Fernandez'
idea of the mode of polyembryonic development in
*D. hybridus*: the entire ectoderm of the individual
embryos originates from the undivided primary ectodermic vesicle through a series of irregular and complicated
outgrowths, while the endoderm and mesoderm of each
embryo is produced *in loco* from the common mesoderm
at points where the outsprouting ectoderm comes in
contact with them. This haphazard method of origin of
the embryos makes it clear then why the number of
embryos in the *Mulita* is not fixed but so highly
variable.

This explanation of polyembryony is purely descriptive and implies no theory as to the factors responsible
for the condition. Some of Fernandez' figures of the
common amnion and the interrelations of the amniotic
connecting canals seem to indicate that not all of the
outgrowths of the ectodermic vesicle are primary, but
that some of them are secondary or even tertiary. In
one case (Fig. 27) there appear to be four independent
primary outgrowths, each connected separately with
the common amnion, and one compound outgrowth,
consisting of four branches, one of which is much
smaller than the rest. In another case (Fig. 30) the
common amnion seems to have divided into two vesicles
united by a narrow neck. From one half-vesicle seven

embryos have arisen; from the other only one rudimentary or degenerate embryo. Other arrangements of embryos are shown in Figs. 28 and 29.

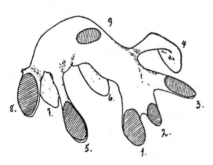

FIG. 27.—Drawing of a wax model of the ectodermic vesicle of an egg of *D. hybridus*, showing the relationship of the nine embryonic outgrowths. Note the great irregularity and lack of definite pairing. (From Fernandez.)

This highly variable condition in the *Mulita* is in sharp contrast with the rather definite one that prevails in *D. novemcinctus*, where about 97 per cent of sets have four fetuses. One cannot help suspecting, however, that, as was previously suggested, the quadruplet condition in the last-named species is not always arrived at in the same way. Just as in the *Mulita* one sprout may remain single and another subdivide once or several times, so in our species it may well be that quadruplets arise, not always in

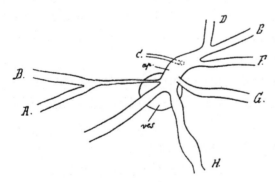

FIG. 28.—Showing the same region of *D. hybridus* that is shown for *D. novemcinctus* in Figs. 21 and 22. Note the very irregular interrelations of the connecting canals of the embryos *A* to *J*. *C* connects with a rudimentary embryo. *A* and *B* might be called a pair; similarly *D* and *E*. (From Fernandez.)

pairs, but sometimes three on one side and one on the other. This would furnish an explanation for the con-

dition that occasionally appears in which three fetuses
are alike and one
quite different.

A striking feature
of Fernandez' collec-
tion of the eggs of
*D. hybridus* has to do
with the frequent,
almost universal,
occurrence of one or
more degenerate
embryos in an egg.
These embryos may
be the victims of
severe competition
for placental surface,
or they may be the
result of outgrowths
produced from un-
favorable regions
of the ectodermic
vesicle. The portion
of an egg shown in
Fig. 31 indicates
that ectodermal out-
growths may fail to
reach the walls of the
egg and, through lack
of endodermic and
mesodermic elements,
may thus fail to be-
come complete em-

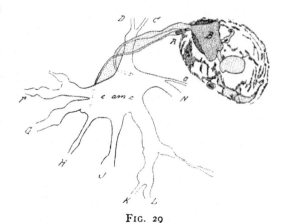

FIG. 29

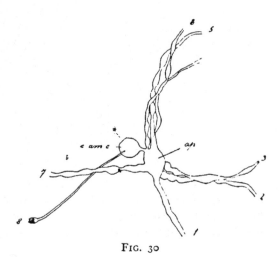

FIG. 30

FIGS. 29 and 30. Two more views of
conditions like those shown in Fig. 28.
There are in the upper figure 13 embryos,
each with a connecting canal; *A*, *B*, and
*C* are evidently rudimentary or aborted
embryos. In the lower figure all of the
successful embryos (1-7) seem to have
come from one-half of the ectodermic
vesicle and only a rudimentary embryo
(8) from the other half. (After
Fernandez.)

bryos. They would also never establish a nutritive connection with the placenta.

A drawing of a wax model (Fig. 27) made from an ectodermic vesicle from which numerous embryonic outgrowths are being given off shows two outgrowths that give promise of rudimentation. Embryonic outgrowth 9 has evidently been produced too high up or too close to the original apical end of the vesicle and

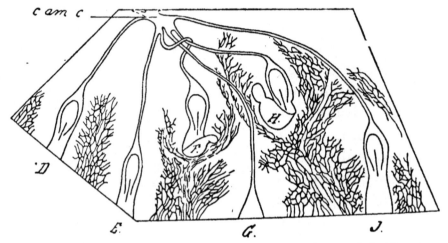

FIG. 31.—A group of primitive streak embryos of *D. hybridus* showing how some of the ectodermic outgrowths (*F* and *H*) fail to secure attachment to the endodermic parts of the egg. This may account for the rudimentary embryos seen in Figs. 29 and 30. (After Fernandez.)

has not been able to grow out as rapidly as the rest; embryo 2 is small and apparently subsidiary to 1 and would probably have been rudimentary. Rudimentary embryos that are no more than formless masses of tissue attached by amniotic connecting canals with the common amnion are shown in Fig. 29. Here also are shown marked irregularities in the interrelations of embryos, as indicated by the forking of amniotic canals.

A good many cases of rudimentary embryos have been found in *D. novemcinctus*, but the mortality of embryos in that species is very low.

## TWINS IN *EUPHRACTUS VILLOSUS*[1]

Late in the year 1915 there appeared a third paper by Fernandez[2] which gives important information about the development of the hairy armadillo or Peludo (*Euphractus villosus*).[3]

*Euphractus* possesses a mode of development strikingly different from that of *Dasypus* and a new and totally different kind of twinning. When one examines the advanced embryonic vesicle of this species, he finds a situation like that shown in Fig. 32. Usually two fetuses are present within what appears to be a single chorion, and they are separated from one another by a membrane that appears to be made up of the fused amnia of the two fetuses. Each fetus has its own separate umbilicus, and the two are not fastened to the wall of the vesicle with any regard to the uterine axes of symmetry. Fernandez examined ten advanced vesicles and found that in seven cases the twin fetuses were of opposite sexes; in two cases both were female and in one case both were male. The occurrence of fetuses of opposite sexes seemed strange in the case of twins that had every appearance of being monochorial; hence the situation deserved careful investigation.

[1] This species has been incorrectly attributed by Fernandez to the genus *Dasypus*.

[2] M. Fernandez, *Anat. Anzeiger*, Bd. 48, No. 13/14, 1915.

[3] The genera *Euphractus* and *Dasypus* are by some authors placed under one genus, and not without some show of reason, for the species of the two genera are certainly very similar.

A considerable collection of 34 embryonic vesicles was assembled, showing stages from the primitive streak stage up to a stage of advancement similar to that shown in Fig. 32.   The net result of the analysis of

FIG. 32.—Photograph (after Fernandez) of a double embryonic vesicle of the armadillo *Euphractus* (*Dasypus*) *villosus*, showing the two fetuses inclosed in what appears to be a common chorion and separated by an amniotic partition composed of the fused amnia of the twins. These come from two eggs and may be or may not be of opposite sex.

this material was that the twins proved *not* to be derived from one egg (i.e., were not monozygotic) at all.   The false appearance of polyembryony was due to the very intimate fusion of two originally quite separate eggs, which were merely crowded so closely in the small, simple uterus that their external membranes fused together so as to simulate a monochorial condition. It is certainly a strange circumstance that within a small group of closely related and sharply differentiated forms like the armadillos there should occur two methods of twinning so diametrically opposite in character: the splitting up, as in *Dasypus*, of a single egg into several embryos, and the intimate fusion, as in *Euphrac tus*, of originally separate eggs into one vesicle. Stranger still is the fact that the end-results of the two processes are so strikingly similar as to lead one to believe that they are the result of the same fundamental processes. Such a finding as this should indicate the necessity of caution in interpreting the various types of monochorial conditions in human twins, for it seems highly probable that the common chorion in many observed cases may, as in *Euphractus*, be due merely to a secondary fusion of originally separate membranes.

Another interesting discovery is that in *Euphractus villosus* occasional cases of uniparous births occur. Fernandez found five such cases in thirty-four pregnant uteri. Three interpretations of this condition are obviously possible. There may be: (*a*) a certain amount of prenatal mortality of one twin of a pair; (*b*) a failure to ovulate on the part of one of the ovaries; or (*c*) one of the eggs may fail to be fertilized. Other reasons might be suggested, but since Fernandez does not furnish the crucial data necessary for deciding the matter—a statement as to whether in these cases of single embryos one or two corpora lutea occur—it would be useless to speculate further. It seems strange that one who knows so well the unique value of the corpus luteum as a means of distinguishing between multiple births and polyembryony should omit to furnish this information.

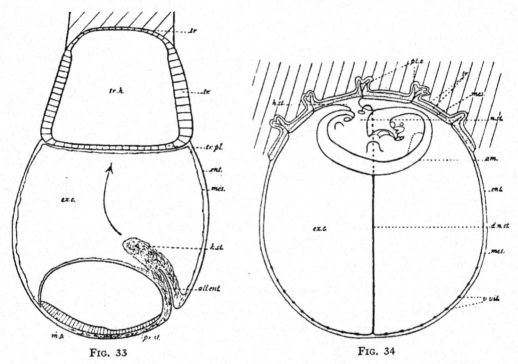

FIG. 33          FIG. 34

FIGS. 33 and 34.—Two stages in the development of the egg of the armadillo *Euphractus* (*Dasypus*) *villosus* (from Fernandez), showing how each of the individuals shown in Fig. 32 appears before they fuse membranes. In Fig. 33 the single embryo has formed at the lower pole, as in *Dasypus* (cf. Figs. 13 and 14), but only one primitive streak arises. In Fig. 34 the single embryo has grown backward, as in *Dasypus*, to the upper pole, where placentation has occurred.

The Peludo shows us typical armadillo development unobscured by polyembryonic processes, and on that account should throw light on the mechanics of polyembryony. A comparison of the earliest known stages of the development of the Peludo and those in *Dasypus* may be made by examining the diagram of Fernandez

(Fig. 33) in connection with Figs. 12 to 14 for *Dasypus*. In both, through the process of so-called germ-layer inversion, the ectodermic vesicle is at the distal pole of the fixed vesicle and the endodermic vesicle covers the external part of the vesicle from the Träger downward. Evidently then this process of germ-layer inversion is not, as was at first supposed, the key to polyembryony. It merely furnishes the conditions under which poly-embryonic development may easily occur. Note that in *Euphractus* the Träger is in the form of a vesicle with a completely inclosed cavity called the Träger cavity. In *D. novemcinctus* there appears to be no true Träger cavity, for the proximal wall, corresponding to that at the top of the Fernandez figures, does not develop. The trophoderm plate is the same in both species, and the thickened ring of trophoderm (the Träger) that in *Dasypus* invades the maternal mucosa in stages represented in Figs. 12 to 16 corresponds to the thickened side walls of the trophodermic vesicle in *Euphractus*. The difference in the two species is associated with the fact that in *Euphractus* the trophodermic vesicle does not sink deeply into the maternal mucosa but lies more on the surface, while in *Dasypus* the penetration of the Träger into the maternal mucosa is much deeper, involving a practically complete dropping of the proximal wall of the Träger cavity. The sides of this vesicle persist as an ingrowing ring or collar of glandular cells that penetrate rather deeply into the mucosa epithelium; but the trophodermic plate inclosed by this ring remains free from the mucosa and may sometimes have between it and the underlying mucosa a considerable space which may be homologized with the Träger cavity of Fernandez.

Although the embryo, now represented mainly by the ectodermic vesicle, at first lies at the distal pole of the egg, it subsequently shifts its position until it appears to be attached to the proximal pole of the egg, as shown in Fernandez' figure. The explanation of this apparent total reversal of position is not far to seek, for there is a close analogy in this respect between *Dasypus* and *Euphractus*. Referring again to Fernandez' figure (Fig. 34), we see that the *Haftstiel* or primitive umbilicus shows the same relation to the endodermal allantois that exists in the genus *Dasypus*. The embryo appears to have grown backward along the vesicle wall through an arc of over 180 degrees. The result is that the anterior end of the embryo which formerly pointed to the left now points to the right; the dorsal aspect, that formerly faced toward the proximal pole, now faces toward the distal pole. Note the slender *Nabelstrang* in Fig. 34, running from the junction of amnion and umbilicus to the distal pole of the vesicle, which, I believe, is the equivalent of the amniotic connecting canals in *Dasypus*. Analysis of the situation reveals the important fact that each quadruplet embryo in *Dasypus* goes through the same reversal of axis, the same backward growth toward the Träger, and the same establishment of a placental connection with the latter as does the single embryo of *Euphractus*. Starting with a single ectodermic vesicle in both species, in *Euphractus* a single apical end and embryonic axis is established, and in *Dasypus* four or more apical ends appear. The real problem of polyembryony is to account for the appearance of four growing points in a vesicle that primitively had but one.

There is evidence that *Euphractus villosus* may be evolving toward a uniparous condition, for single fetuses occur in about 15 per cent of the cases observed by Fernandez. This author also observes that the uterus of a young female is distinctly bicornate in structure, a fact that may serve as evidence that multiple gestation was the primitive condition and that the occurrence of a single birth is a modern tendency resulting from the gradual transformation of the uterus from the primitive bicornate form into the simple form.

Two closely allied species of *Dasypus*, *D. sexcinctus* (the six-banded armadillo) and *D. gymnarus* (the one-banded armadillo), have been described by von Kölliker and by Chapman as normally uniparous, but Carson in a letter states that he had a female *D. sexcinctus* in the Philadelphia Zoölogical Garden that in three successive pregnancies produced twins twice and a single fetus once. Evidently then twinning is quite common among the armadillos and is probably the normal condition in many. In the little three-banded armadillo, *Tolypeutes conurus*, the uniparous condition is evidently typical, as Fernandez says of it: "Alle trachtigen uteri des *Mataco* (*Tolypeutes conurus*) führen nur einen Embryo." He does not state, however, upon how many cases this statement is based.

One may infer then that the earliest ancestors of the modern armadillos had bicornate uteri and had multiple offspring; that the next step was a shortening of the horns and a tendency to produce twins; that with the development of a simplex uterus there came a tendency to produce single offspring, which, in some species has become a specific character. In the genus

*Dasypus*, which is evidently the most highly specialized genus of the Dasypodidae, the process of polyembryony is the last step and has been derived from a uniparous situation through the introduction of some factor that causes the single blastocyst to undergo precocious agamic reproduction. Polyembryony is, therefore, not a primitive but a highly specialized condition, though some authors refer it back to conditions in the lower chordates or even to conditions in the invertebrates.

# CHAPTER IV

## THEORIES OF POLYEMBRYONIC DEVELOPMENT IN DASYPUS

A clue to a physiological explanation of polyembryony appears to be presented by a comparison of the developmental rates of uniparous and of polyembryonic armadillos. In both species of *Dasypus* the gestation period is abnormally long for a mammal of such comparatively small size, covering a period of from four to five months; in *Euphractus villosus*, in which one egg gives rise to but one embryo, the period of gestation is only two months. So short is the gestation period that two broods are produced in a season instead of one, as in *Dasypus*. Fernandez attempts to explain the slower development of *Dasypus* by saying that the nutriment received by the embryos from the maternal blood is not sufficient to allow development to proceed at the accustomed rate, and that the crowding of many fetuses in a simple uterus tends further to retard development. He forgets, however, that polyembryonic development begins and is fully completed long before the young egg establishes a permanent nutritive relation with the maternal tissues and that there is no crowding in the uterus until the completely formed embryos have attained a considerable degree of advancement. We cannot, therefore, explain the establishment of four or more growing points at the apex of the ectodermic vesicle as due to insufficient maternal nutriment or to crowding.

A much more satisfactory explanation is associated with the fact that there is an early "period of quiescence" of the egg.   This fact, though no significance was attributed to it, was brought out by Patterson, who found that all of the eggs collected over a period extending from the middle of October to the fourth of November were in almost the same stage of development.   It was found, moreover, that eggs in the Fallopian tubes were almost as advanced as those found practically in the area of placentation.   These observations prove that the factors responsible for retardation of development in the polyembryonic species are in operation at a very early period, probably as early as the first cleavage stages.   The problem is to locate the factors responsible for the slowing down of the developmental rhythm. Whatever these factors may be, and we have no definite knowledge of them, the result of retardation is polyembryony.

For some years I have been convinced that this case of agamic reproduction is physiologically equivalent, in some important respects, to the well-known case of budding in plants.   This view was expressed in 1913 in a general paper[1] on the natural history of the nine-banded armadillo.   In that paper I ventured, perhaps unwisely, to attribute the retarded development to the presence of a specific parasite in the armadillo egg.   This structure had at that time been diagnosed for me by an expert protozoölogist as a sporozoan parasite.   At the present time I am unable to decide whether the object in question is a protozoan or not; at least it is of universal occurrence in the armadillo egg,

[1] H. H. Newman, *American Naturalist*, XLVII, 1913.

and is not found in the eggs of other mammals.   Whether the structure is the cause of retarded development, or merely one of its results, cannot be decided at present. . In the paper just referred to I made the following statement, which deserved, I believe, further elaboration:

According to Professor Child's theories of development and reproduction, any part of a system, which, through a lowering of the rate of metabolism of the controlling part of the system, say the animal pole of the blastodermic vesicle, is liable to physiological isolation of [subordinate[1]] parts at certain distances from the dominant region.   When such isolation of [subordinate] parts occurs, new centers of control arise, which produce outgrowths [apical ends] capable of establishing new systems like the original.

A familiar instance of agamic reproduction in plants will serve to make this theory clear.   In a plant the growing tip (apical end) is the dominant end of the branch and it seems to hold in physiological subordination a considerable part of the branch distal to itself. If, however, anything happens to this growing tip (apical end) which lowers its rate of metabolism, a whirl of secondary growing tips will appear just back of the original apical end, and each of these new apical ends will become new dominant regions, capable of producing new individuals.[2]   Any type of environmental change that has the effect of lowering the rate of metabolism of the plant will have the effect of

---

[1] The bracketed words in this quotation were not in the original but are inserted here for the sake of clearness.

[2] In plants the individual is not so sharply defined as in most animals.   For our purposes we may consider each growing point an individual and the whole plant a colony.

inhibiting the dominance of the apical end and of producing a whirl of secondary growing points.

Now in the armadillo egg the ectodermic vesicle has an apical point, which is the head end or growing tip (see Fig. 12, X) of the embryo before the process of fission occurs. If the conditions of growth were such as to admit of a normal rate of metabolism, this original apical end would become the head end of a single embryo. Some agency lowers the rate of metabolism of the embryo and the original apical end loses its dominance over subordinate regions; the result is that several radially arranged secondary points in the ectodermic vesicle acquire independence. Those that are most favorably situated with reference to the uterine axes express their independence first and become the first visible growing points, the so-called "primary buds"; those that are less favorably situated acquire independence a little later and form the so-called "secondary buds." It happens that, almost synchronously with the physiological isolation of the whirl of subordinate growing points, a new and effective nutritive connection (the Träger ring) is established between the embryonic vesicle and the maternal tissues, which greatly accelerates the metabolic rate and the consequent speed of growth. This rejuvenating factor stops the production of further growing points and makes it possible for each of the newly formed apical ends (heads) to develop a body. When the conditions of growth are restored to normal, the vesicle is no longer a single individual but is a clone, consisting of four essentially separate individuals, each of which goes through its own embryonic development quite independently of the others, except in so far as development within a common chorion and the necessity of sharing a single primary placenta involve mutual adjustments.

Various other theories purporting to offer explanations of the production of plural embryos from single eggs have been advanced, but the three that have been most persistently maintained are the "blastotomy theory," the "budding theory," and the "fission theory."

BLASTOTOMY VERSUS BUDDING

Is each embryo the lineal descendant of a single blastomere of the four-cell stage of cleavage or does the process of fission heretofore described occur without reference to the cell products of the early cleavage cells? About these contrasting theories of polyembryony in the armadillo there has been much difference of opinion and some shifting of viewpoints on the part of individual workers. In one of their earlier papers (1910) Newman and Patterson stated that it seems highly probable that the tissues involved in each of the four quadrants of an embryonic vesicle do really arise as the lineal descendants of one of the first four blastomeres. Still earlier in 1909 they said: "In the case of *Dasypus* each embryo probably arises from one of the blastomeres of the four-celled stage." It may be said that, so far as the writer is concerned, no idea of blastotomy in the sense that there was any actual physical separation or isolation of blastomeres was ever entertained. Patterson, however, in 1913, in discussing the various theories of polyembryony, classes the theory "that each embryo is the lineal descendant of a single blastomere of the four-celled stage" as "spontaneous blastotomy." The very meaning of this term involves the idea of physical separation of blastomeres and is therefore quite inappropriate in connection with the idea that had been expressed by the joint authors. I quite agree that no true "blastotomy" occurs, but I would maintain that there is much evidence for and little against the idea that the cleavage process of the armadillo is determinate, that the cell descendants of each blastomere of the four-cell stage constitute essentially a quadrant of the vesicle (including trophoblast, ectoderm, endoderm, etc.), and that therefore the inherited characters of each embryo are dependent upon the particular quadrant, or parts of different quadrants, from which it is derived.

Patterson, however, totally abandons the idea, formerly entertained at least tacitly by him, that there is any connection between the four embryos and the four blastomeres. This change of opinion is doubtless due to the observation that budding appears to occur in accordance with the bilaterality of the uterus rather than in accordance with any lines of demarcation predetermined in the egg.

In a recent paper Wilder (1916), in discussing the armadillo results, continues to maintain the view that there is a real connection between the four blastomeres and the four fetuses. He says that it is now clearly evident that all ideas of physical separation of these blastomeres, a definite *blastotomy*, does not take place, yet many things still point to the conclusion that a similar condition is obtained through some form of differentiation, and that each of the separate embryonal *anlages*, be they two, or four, eight, or eleven (a possible number in *Tatusia hybrida*[1]), is the lineal descendant of a single blastomere, formed during early cleavage.

## THE FISSION THEORY

In a paper read before the Ninth Zoölogical Congress at Monaco, Assheton criticizes both the "blastotomy" and the "budding" theories of Newman and Patterson. The theory of budding seems to him especially unacceptable. "One cannot have budding," he says, "unless there is a stock from which budding takes place. There is nothing in *Tatusia* [*Dasypus*] one can call a stock. The phenomenon is clearly that of fission."

In support of the fission hypothesis, he cites evidence derived from the embryological study of other mammals, notably the sheep and the ferret. "In the sheep [*Ovis*] I found some years ago a blastocyst at the stage just before the formation of the embryonal areas with two distinct ectodermic masses lying within the trophoblast." His outline figures show this interesting sheep blastocyst (Figs. 35 and 36) to be covered by a complete envelope of trophoblast and lined internally with a complete layer of endoderm. Such a condition could not have resulted from budding.

In the ferret (*Putorius*) certain interesting conditions (Figs. 37 and 38) were found that seemed to show evidences of a separation of blastomeres, but in no case were twin embryos produced. These cases are cited to prove "that fission of the embryonic rudiments of eutherian mammals may be effected fairly easily, but the occurrence is the exception, not the rule." A

[1] Meaning *Dasypus hybridus*.

constructive theory of polyembryony in *Tatusia* (*Dasypus*) is then offered, which is based upon the unique combination of three conditions:

(1) the development of the blastocyst within the central lumen of the uterus which has allowed of a considerable expansion of

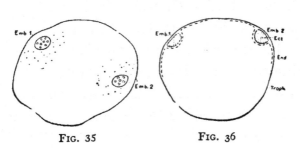

the ectodermic plate, owing to the rolling up of the blastocyst cavity as in *Lupus* now; (2) 'inversion of layers' by which the ectoderm plate becomes invaginated into the large cavity of the blastocyst subsequent to its expansion; (3) a late formation of a thickened

FIG. 35          FIG. 36

FIGS. 35 and 36.—Two views of sheep ovum with twin embryo (after Assheton). It is unlikely that such double embryos develop very far.

ened mass of trophoblast over the entire expanded plate putting more pressure on the center than on the periphery of the ectodermal disk.

Assheton points out that "if we take a case like that of *Lupus* and superimpose upon it the Träger of *Mus* we should get a condition which would approximately be that of *Tatusia* [*Dasypus*]."

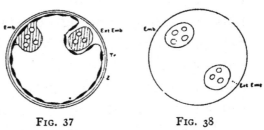

FIG. 37          FIG. 38

FIGS. 37 and 38.—Two views of a double embryo of the ferret (after Assheton).

This purely morphological explanation of what seems to me unquestionably a physiological process serves only to obscure the real problem of the causal basis of polyembryony. I am, however, in agreement with Assheton in his objection to the budding hypothesis and in considering the process by which the individual embryonic primordia emancipate themselves from the common germ as a process of *fission*, or dividing up of the ectodermic vesicle into several ectodermic primordia, each of which stands for an embryonic area. The part that is left behind, the common amnion, is a comparatively insignificant residue and could scarcely be termed the stock.

CONCLUSION

In reviewing all of the facts it now seems to me that the position that there is a genetic connection between the four embryos and the four blastomeres is entirely untenable. The extraordinarily irregular arrangement and number of fetuses in *D. hybridus* seem to be totally out of accord with this idea. The budding theory, though affording a convenient descriptive device, appears to be open to serious objection, as shown by Assheton, yet the mechanism of outgrowth production is not unlike certain true cases of budding. The fission idea seems to be on the whole less open to criticism, if by fission we mean merely the physiological isolation of several secondary points in a single embryonic vesicle, and the consequent acquisition by these points of independence in growth and development.

Unquestionably long before the isolation of several secondary growing points a considerable amount of differentiation has occurred, so that genetic factors are unequally distributed in the various regions which give rise to the new apical points. We may expect a less degree of difference between closely adjacent points than between widely separated points; hence the phenomenon of closer resemblance between pairs derived from one-half of the egg. Evidently, too, as is brought out better in another connection, certain symmetry relations of the entire egg are established before the formation of secondary growing points, and the residuum of this early symmetric individuation is expressed in mirror-image resemblances among the quadruplet fetuses.

# CHAPTER V

## TWINNING IN RUMINANTS—THE FREEMARTIN

All ruminants produce habitually but one offspring at a birth, but in several species (probably in all) two or more offspring occasionally are born at once. Such offspring are naturally spoken of as twins or triplets. Twinning in cattle and in sheep has received considerable attention because of the economic aspects of the case; if it should prove to be an inherited trait, it would be a distinctly advantageous character to encourage by breeding.

Less attention has been paid to twinning in wild ungulates; that the phenomenon occurs in the deer is proved by the beautiful picture of a doe with twin fawns published some years ago in the National Geographic Magazine.[1] The twin fawns are remarkably alike, and if one were to judge by appearances alone, he might be inclined to class them as monozygotic or duplicate twins. Such unsupported evidence would, however, scarcely justify the conclusion.

At the present time I have no reliable evidence of twinning in horses, but it is highly probable that this group offers no exception to the general rule that all mammals normally producing but a single offspring at a birth may have twins. There are certain evidences of a kind of twinning in swine involving the formation of double monsters. That separate monozygotic twins

[1] *National Geographic Magazine*, XXIV (1913), 762.

occur, however, seems rather improbable, from our knowledge of the early embryology of swine. The field needs further investigation.

*One fact about the twins of both cattle and sheep which places them in a different category from armadillo quadruplets is that the twins may either both be of the same sex or of opposite sexes.* If sex is determined at the time of fertilization, it seems unlikely that twins of opposite sexes could come from one egg.

Bovine twins may be of four types: (*a*) two normal males; (*b*) two normal females; (*c*) a normal male co-twin with a normal female; (*d*) a normal male with a *freemartin*. The exact nature of the freemartin is a question about which there has been great diversity of opinion; it is the freemartin situation which lends special interest to the study of twinning in cattle.

### THE PROBLEM OF SEX IN THE FREEMARTIN

The freemartin is a sterile twin born co-twin to a normal male. This definition is quite free from implication as to the sex of the sterile individual, and advisedly excludes the so-called "fertile freemartin."

John Hunter[1] (1786) appears to have been the first to study the freemartin and to report an opinion as to its nature. He described three specimens and diagnosed the conditions as follows: The first specimen (seven years old) was apparently a *hermaphrodite* in that it had a vagina, a rudimentary bicornuate uterus, testes, and vasa deferentia and vesiculae seminales; the second (five years old) was "more like an ox or spayed heifer" in general appearance. It had a vagina with

[1] J. Hunter, London, 1786.

a blind end, a two-horned uterus, testicles nearly as large as those of a bull, and no ovaries. Seminal vesicles opened into the vagina, but there were no vasa deferentia. A clitoris was present. This animal was preponderatingly male, but showed organs of both sexes. The third (three or four years old) was more like a heifer. There was the beginning of a vagina open as far as the urethra, a two-horned uterus, and paired ovaries. But there were also seminal vesicles and part of the vasa deferentia. This individual was predominantly female, but also had organs of both sexes (a hermaphrodite). Hunter's view that the freemartin is a transverse hermaphrodite with a varying predominance of the two sexes is the classical one, and it went unchallenged for some time.

The most extensive collection of data up to that of Lillie was made by Numan,[1] a Dutch investigator, whose paper, together with an atlas of illustrations, I have had the opportunity of looking over. This monograph has been translated into French and from the French into German by Spiegelberg. In the course of these translations some accuracy has been lost. Hart gives what proves to be a very poor and inaccurate translation of a summary of Numan's findings. In brief, it may be said that Numan claims to have evidence of the following kinds of opposite-sexed twins in cattle: (1) normal male with sterile female (freemartin); this is much the commonest type; (2) normal male with normal female; this is a rare type; (3) normal female with sterile male; this is an extremely rare type and is based on second-hand or hearsay information. Numan pictures both

[1] A. Numan. Utrecht: Van der Monde, 1843.

the external and internal genitalia of a considerable number of freemartins and shows clearly that they range from only slightly abnormal female types, in which the development of the female organs is merely retarded or juvenile in condition, up to those which appear to have developed certain positive male characters, such as testes. Numan's data are extensive and valuable, but his interpretations are open to question.

Spiegelberg[1] (1861) was the next investigator to take up the problem of the freemartin. After a careful study of Hunter's and Numan's works he secured and examined on his own account two cases of full-term twins, giving a detailed description of the gross and microscopic anatomy of all significant parts. His conclusion was that the freemartin is not an imperfect or sterile female, but an imperfect male.

Hart[2] gives a summary of freemartin literature, drawing largely from Numan's and Spiegelberg's data. His own conclusions are totally erroneous. He was able to make microscopic sections of the gonads of Hunter's freemartins that had been preserved in the British Museum. In each case they appeared to him to have the histological structure of testicular tissue. On the basis of this evidence, taken with other data previously presented, he says: "It seems to me, therefore, fully established that the freemartin, when the co-twin of a potent male, is a sterile *male* and not a sterile female: i.e., they are identical twins except in their genital tract and secondary sexual characters."

---

[1] O. Spiegelberg, *Ztsch. für rationalle Medicin*, Henle and Pfeufer, Drt. Reihe, Bd. XI (1861).

[2] D. Berry Hart, *Proc. Royal Soc. Edin.*, XXX (1910).

From this and other statements it is clear that Hart
considered the freemartin and its twin male as deriva-
tives of a single fertilized egg (monozygotic). On this
assumption, which is not well founded, as we shall see
later, he builds up a hypothesis to explain how the
freemartin arises, which may be briefly summed up as
follows: Thus the freemartin with a potent bull-twin is
the result of the division of a male zygote, so that the
somatic determinants are equally divided, the genital
determinants unequally divided, the potent going to
the one twin, the potent bull, the non-potent genital
determinants going to the freemartin. The potent
organs are dominant, the non-potent recessive. The
Mendelian scheme may be figured as follows:

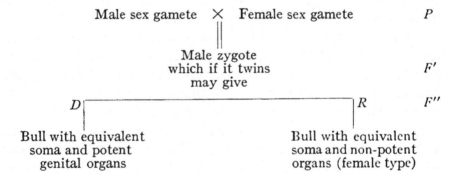

This Mendelian view of the freemartin as a pure
or extracted recessive lacking its genital determinants
is one of the oddest developments of the neo-Mendelian
cult and might be discussed pro and con at some
length were it not for the fact that the whole hypothesis
is based on an error, for *bovine twins are not monozygotic.*

It will be noted that while the earlier workers considered the freemartin as a hermaphrodite, Hart considers it a recessive or non-potent male, co-zygotic with

a potent male.  This interpretation is evidently due to some extent to the preconceived idea that the twins are monozygotic and therefore should be of the same sex. It would seem unlikely that twins of opposite sex are ever derived from a single fertilized egg.  Such a finding would be totally out of accord with what we know of the chromosomal basis of sex-determination in mammals.  Yet such a view has not been without adherents.  Bateson, for example, in his *Problems of Genetics* makes the following statement about free-martins:

> In horned cattle twin births are rare, and when types of twins of *opposite sexes* are born, the male is perfect and normal, but the reproductive organs of the *female* [italics mine] are deformed and sterile, being known as a freemartin.  The same thing occasionally occurs in sheep, suggesting that in sheep also twins may be formed by the division of *one ovum* [italics mine]; for it is impossible to suppose that mere development in juxta-position can produce a change of this character.  I mention the freemartin because it raises a question of absorbing interest.  It is conceivable that we should interpret it by reference to the phenomenon of gynandromorphism, seen occasionally in insects, and also in birds as a great rarity.  In the gynandromorph one side of the body is male, the other female.  In such cases neither side is sexually perfect.  If the halves of such a gynandromorph came apart, perhaps one would be a freemartin.

This statement commits Bateson to the theory of monozygotic origin of heterosexual cattle and sheep twins and to the interpretation of the freemartin as a sterile female.

Recently Cole[1] in a brief abstract takes a view of the value of the freemartin quite in accord with that

[1] L. J. Cole, *Science*, N.S., XLIII (1916).

of Hart, and cites certain statistical evidence in favor of it. As the abstract is so brief it may be quoted in full:

A study of 303 multiple births in cattle, obtained directly from the breeders. The records include: 43 cases homosexual male, 165 cases recorded heterosexual (male and female), 88 cases homosexual female, 7 cases triplets, a ratio of twins approximately 1 : 4 : 2 instead of 1 : 2 : 1 expected, if there were no disturbing element entering in. The expectation may be brought more nearly into harmony with the facts if it is assumed that in addition to ordinary fraternal (dizygotic) twins there are numbers of "identical" (monozygotic) twins of both sexes, and that while in the case of females those are both normal, in the cases of a dividing male zygote, to form two individuals, in one of them the sexual organs remain in the undifferentiated stage, so that the animal superficially resembles a female and is ordinarily recorded as such, although it is barren. The records of monozygotic twins accordingly go to increase the homosexual female and the heterosexual classes, while the homosexual male class, in which part of them really belong, does not receive any increment. This brings the expected ratio much nearer the ratio obtained.

Any female calf twinned with a male is referred to as a freemartin. According to the interpretation given, some freemartins should be fertile while others are sterile. It was found that both exist.

It will be noted that Cole interprets the freemartin as an undeveloped male in which the sex-organs remain in the undifferentiated state and thus resemble those of a juvenile female. This view of the freemartin as a male is a concession to the idea that monozygotic twins should be of the same sex, since sex is supposed to be determined at the time of fertilization.

As is always the case when expectations are based on incomplete data, these numerous divergent interpretations as to the nature and mode of origin of the freemartin are all quite mistaken; now that we have the problem really cleared up they seem almost equally absurd. The Mendelian interpretation of Hart, the suggestion involving gynandromorphism of Bateson, and the inferences of Cole based on sex-ratios appear alike far-fetched in the light of further facts and more reliable conclusions.

Recently F. R. Lillie[1] has solved the mystery of the freemartin through an embryological study of twinning in cattle. The material has been collected during the past two or three years, and I have been much interested in the progress of the research. Lillie's preliminary paper was called forth as a reply to Cole's report just given; he criticizes Cole's data and his interpretations, and says:

I wish to point out the fatal objection that, according to the hypothesis, the females remaining in the heterosexual class are normal; in other words, on this hypothesis, the ratio of normal freemartins (females co-twin with a bull) to sterile ones is 3 : 1; and the ratio would not be very different on any basis of division of the heterosexual class that would help out the sex-ratio. Hitherto there have been no data from which the ratio of normal to sterile freemartins could be computed, and Cole furnishes none. I have records of 21 cases statistically homogeneous, three of which are normal and 18 abnormal. That is, the ratio of normal to sterile freemartins is 1 : 6 instead of 3 : 1.

This ratio is not more adverse to the normals than might be anticipated, for breeders' associations will not register freemartins until they have proved capable of breeding, and some breeders hardly believe in the existence of fertile freemartins, so rare are they.

[1] F. R. Lillie, *Science*, N.S., XLIII (1916).

Having shown the untenability of Cole's conclusions, Lillie presents the results of his own examination of 41 cases of bovine twins, all examined *in utero*.
The material was collected from the Chicago stockyards
by a skilled assistant, and in every case the ovaries of
the mother were obtained and the embryonic envelopes
of the fetuses were preserved and examined. The
nature of the fetal genitalia was determined by dissec-
tion, and in some cases by sectioning. No such body
of data has previously been obtained on bovine twins.
This work is still in progress and will be reported at
length in due time. In the meantime it will be well to
give here nearly all of the data furnished by Lillie's
abstract in *Science*.

Out of 41 cases of bovine twins 14 are both male,
21 are of opposite sexes, and 6 are both female. This
is about what one would expect, if we interpret the freemartin as a female, for there are about as many same-
sexed twins as opposite-sexed. The number of the
same-sexed male twins is higher than expected, but
perhaps this is due to the comparatively small number
of cases. If, according to Hart and Cole, freemartins
also are males, there would be an enormous and inexpli-
cable preponderance of males. As Lillie says:

The real test of the theory must come from the embryological
side. If the sterile freemartin and its bull-mate are monozygotic,
they should be included in a single chorion, and there should be
a single corpus luteum present. If they are dizygotic, we might
expect two separate chorions and two corpora lutea. The
monochorial condition would not, however, be a conclusive test
of monozygotic origin, for two chorions, originally independent,
might fuse secondarily. The facts as determined from examination of 41 cases are that about 97.5 per cent of bovine twins are

monochorial, but in spite of this nearly all are *dizygotic* [italics mine]; for in all cases in which the ovaries were present with the uterus *a corpus luteum was present in each ovary* [italics mine]; in normal single pregnancies in cattle there is never more than one corpus luteum present. There was one homosexual case (males) in which only one ovary was present with the uterus when received, and it contained no corpus luteum. This was probably monozygotic.

In cattle a twin pregnancy is almost always the result of the fertilization of an ovum from each ovary; the development begins separately in each horn of the uterus. The rapidly elongating ova meet and fuse in the small body of the uterus at the same time between the 10 mm. and 20 mm. stage. The blood vessels from each side then anastomose in the connecting part of the chorion; a particularly wide anastomosis develops, so that either fetus can be injected from the other. The arterial circulation of each overlaps the venous territory of the other, so that constant interchange of blood takes place [Fig. 40]. If both are males or both are females, no harm results from this; *but if one is male and the other female, the reproductive system of the female is largely suppressed, and certain male organs even develop in the female.* This is unquestionably to be interpreted as a case of hormone action. It is not yet determined whether the invariable result of sterilization of the female at the expense of the male is due to more precocious development of the male hormones, or to a certain natural dominance of male over female hormones

The results are analogous to Steinach's feminization of male rats and masculinization of female by heterosexual transplantation of gonads into castrated infantile specimens. But they are more extensive in many respects because of the incomparably earlier onset of the hormone action. In the case of the freemartin, nature has performed an experiment of surpassing interest.

Bateson states that sterile freemartins are found also in sheep, but rarely. In the four twin pregnancies of sheep that I have so far had the opportunity to examine, a monochorial condition was found, though the fetuses were dizygotic; but the circulation of each fetus was closed. This appears to be the normal condition in sheep; but if the two circulations should anastomose, we should have the conditions that produce a sterile freemartin in cattle. The possibility of their occurrence in sheep is therefore given.

The fertile freemartin in cattle may be due to causes similar to those normal for sheep. Unfortunately, when the first two cases of normal cattle freemartins that I have recorded came under observation, I was not yet aware of the significance of the membrane relations, and the circulation was not studied. But I have in my notebook in each case that the connecting link of the two halves of the chorion was narrow, and this is significant. In the third case the two chorions were entirely unfused; this constitutes an *experimentum crucis*. The male was 10.4 cm. long; the female, 10.2 cm. The reproductive organs of both were entirely normal. The occurrence of the fertile freemartin is therefore satisfactorily explained.

The sterile freemartin enables us to distinguish between the effect of the zygotic sex-determining factor in mammals and the hormone sex-differentiating factors. The female is sterilized at the very beginning of sex-differentiation, or before any morphological evidences are apparent, and the male hormones circulate in its blood for a long period thereafter. But in spite of this, the reproductive system is, for the most part, of the female type, though greatly reduced. The gonad is the part most affected; so much so, that most authors have interpreted it as a testis; a gubernaculum of the male type also develops, but no scrotal sacs. The ducts are distinctly of the female type much reduced, and the phallus and mammary glands are definitely female. The genital somatic habitus inclines toward the male side. Male hormones circulating in the blood of an individual zygotically female have a definitely limited influence even though the action exists from the beginning of morphological sex-differentiation.

The drawing of twin calves shown in Fig. 39 shows very clearly the intra-uterine relations of bovine twins. For permission to use this figure I am much indebted to Professor Lillie, whose monograph on cattle twins is now in course of preparation. The figure shows twins

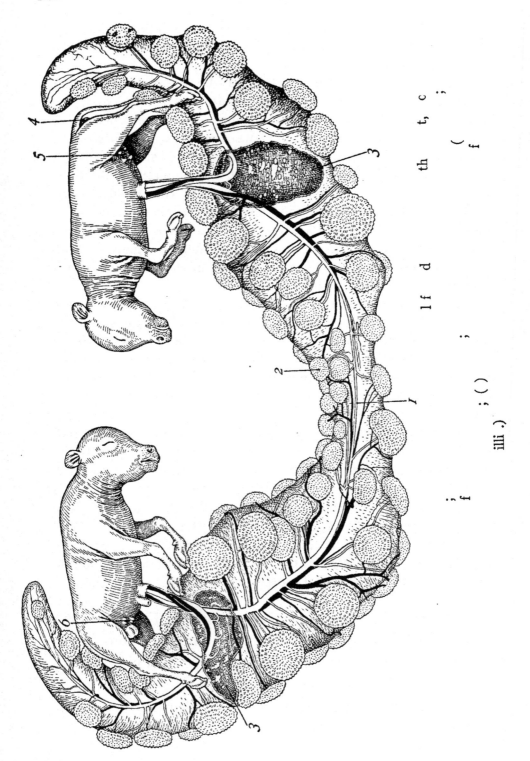

about ten inches long taken out of the chorionic sac through the openings shown in dark shading. The double chorionic vesicle is shown somewhat collapsed through loss of its fluid contents. The narrow part in the middle is the point where two separate eggs have fused at a much earlier period. The flower-like pads scattered over the membranous wall are the placental areas, the so-called cotyledons, by means of which the fetuses obtain nourishment from the uterine tissues. The chorionic blood vessels and those of the fetuses are shown clearly, the arteries in outline and the veins black. It will be seen that the arteries of the two fetuses are in direct communication across the placental bridge. The veins of both fetuses in some cases enter the same cotyledon. There is every opportunity for the admixture of blood between the twin individuals. It need hardly be pointed out that the individual on the left is a male and that on the right a sterile female or freemartin.

Lillie's work has revealed the true nature of the freemartin; it is a sterile female whose gonads remain in the juvenile stage so that they resemble testes, and which has certain secondary sexual characters of the male due to the presence for a considerable period of male hormones in the blood borrowed from its male co-twin. The animal is a hermaphrodite only in a very limited sense. The work leaves no question as to the dizygotic origin, not only of opposite-sexed, but also of same-sexed bovine twins.

Whether real monozygotic twinning ever occurs among the ungulates is highly questionable. Several considerations lead to this conclusion. Polyembryony, in those mammals that unquestionably show it, has a definite relation to certain peculiar types of simple uteri and a special type of embryonic development called germ-layer inversion. In the armadillo it is difficult to imagine how, apart from germ-layer inversion, the complete separation of embryos could occur and the monochorial condition be still maintained. In man the conditions described for the Bryce and Teacher ovum appear to indicate a situation like that of the armadillo; but no such condition has been described for any of the ungulates. It is barely possible that blastomeres could separate and form two individual embryos, but this would involve a subsequent fusion of chorions just like that which occurs in the dizygotic twins of *Euphractus villosus*. Lillie cited one equivocal case of apparent monozygotic twins. In one pair of twin males the genitalia of the mother were roughly handled so that when examined but one ovary was present. This ovary had no corpus luteum. The inference made is that the lost ovary had but one corpus luteum, since in all other cases examined but one corpus luteum occurs to an ovary. It seems probable, however, that the lost ovary would have shown two corpora lutea had it been examined. That an ovary may ovulate two ova at a time is seen from the fact that triplets, which may be either of the same sex or of opposite sexes, not infrequently occur. That calf twins may be monozygotic is believed by several writers. Pearl, for example, in a paper on "Triplet Calves" states that the two nearly identical females (see two outer calves in Fig. 40) are possibly, if not probably, derived from the first two blastomeres of a single dividing egg.

This condition seems to me barely possible, but very improbable.

Twinning in the genus *Dasypus* and twinning in ruminants turn out to be two totally different phenomena. In *Dasypus*, the twins are monozygotic and always of the same sex.  In cattle, twins are dizygotic

FIG. 40.—Photograph of triplet calves (from Pearl)

and may or may not be of the same sex.  The additional phenomenon of freemartinism adds materially to the value of bovine twins, especially in its bearings on the problems of sex-biology.  In a subsequent chapter some of these bearings receive a more adequate discussion in connection with other sex-phenomena in twins than is possible at this time.

# CHAPTER VI

## TWINS IN RELATION TO GENERAL BIOLOGICAL PROBLEMS

The study of twins throws light especially on the following problems: (1) the time of and the mechanics of sex-determination; (2) the significance of sex-ratios; (3) the mechanism of sex-differentiation; (4) is twinning hereditary? (5) the modes of inheritance in monozygotic or polyembryonic twins; (6) the nature and significance of symmetry reversals in monozygotic twins. The first five problems are discussed in the present chapter, while the last two problems are reserved for the next chapter.

### TWINS AND THE PROBLEM OF SEX-DETERMINATION

The problems of sex are today attracting the widest attention, and among these problems that of the mechanism of sex-determination appears to have been largely solved. It appears that in a vast number of animals of all grades of organization, from worms to man, sex is determined at the time of fertilization. In some forms sex is determined in the egg, for there are two distinct types of eggs, male-producing and female-producing. In other cases the eggs are all alike and produce females if allowed to develop parthenogenetically (without fertilization), but produce half males and half females if fertilized, the result being due to two kinds of spermatozoa, male-producing and female-producing. In still other cases, notably the hymenoptera, males are produced when the eggs are not fertilized and females when the eggs are fertilized. All of these apparently divergent phenomena are consistent with the idea that sex is determined in the germ-cell and that the sex-determining factor is in some way intimately associated with the presence of a peculiar chromosome (the $X$ chromosome), or group of chromosomes, in the nucleus of the germ-cell. This mechanism gives a sex-bias to the individual, a bias in some cases so strong that no known factors can interfere with the fulfilment of the sex-development that was originally determined. In other cases, however, sex may be zygotically determined, but requires a definite favorable environment to bring it to complete development or differentiation. Finally, in some cases the individual which is zygotically sex-determined may have its sex-development so altered as to become largely of the opposite sex.

In mammals there is much evidence that sex is zygotically determined; there appear to be two kinds of spermatozoa and but one kind of egg, and the sex of the individual depends on whether a male-producing or a female-producing spermatozoön fertilizes a particular egg.

If then sex in mammals is determined in the undeveloped egg, we should naturally expect twins or large numbers of individuals derived from a single egg to be of the same sex. This is just the point upon which monozygotic twinning and polyembryony have a bearing, for in these phenomena we have experiments demonstrating the theory of zygotic sex-determination.

If we were able artificially to subdivide a fertilized egg into two or more parts and the individuals developing from the parts of a given egg were always of the same sex, we should consider the theory of zygotic sex-determination as proven. Our nicest technique, however, has not been adequate to carry out so crucial an experiment, and we are therefore forced to rely upon an equivalent experiment in nature. It has been shown conclusively for two species of armadillo, and by analogy for man, that an egg, divided at an early period into two or more embryonic primordia, produces individuals all of the same sex. In hundreds of sets of quadruplets in the Texas armadillo there has occurred no exception to this rule, in spite of the fact that in some cases there are marked differences in size due to unequal environment factors. In one case two fetuses of a set are nearly twice the size of two others, yet the sex of all is the same, showing that the zygotically determined sex is incapable of alteration through any ordinary environmental change. Similarly, in man twins that are monochorial and in other ways bear evidences of monozygotic origin are always of the same sex. Conjoined twins, which are unquestionably monozygotic, are also same-sexed, although one half of an individual may be much better developed than the other.

Not only in mammals but also in other animals exhibiting polyembryony it is true that all individuals derived from a single germ-cell are same-sexed; several investigators have shown this for various genera of parasitic hymenoptera. Silvestri was the first author to discover polyembryony in these animals. He found that in the genus *Litomastix*, in which the eggs are laid in the body of a caterpillar, a single egg divides very early into a very large number of separate embryonic primordia, each of which produces an adult insect. No matter how many individuals are derived from one egg—the number may be even a thousand—they are of the same sex. Some get more food than others, grow faster, and become larger, but nothing environmental affects the zygotically determined sex. The mode of zygotic sex-determination is somewhat peculiar in that eggs develop whether fertilized or not; if fertilized all offspring from that egg are females; if not fertilized, but developing parthenogenetically, all offspring from a given egg are males.

If then in two groups as far apart as mammals and hymenoptera the fact of zygotic sex-determination is proved by the phenomenon of polyembryony, it seems probable that this is a very general principle; possibly universal.

### SEX-RATIOS IN TWINS AND MULTIPLE BIRTHS

Certain facts derived from a statistical study of the sexes of twins of various species seem at first sight to be opposed to the theory of rigid zygotic sex-determination in mammals. It would appear that the proportion of males to females in twins and multiple offspring, whether monozygotic or dizygotic, should not differ from that which holds for the species in general. In mammals the ratio of males to females is in general quite close to 1 : 1, although a slight preponderance of males is usually present.

In man, for example, it has been shown by several writers that as the number of individuals to a birth increases the relative proportion of males to females decreases.   Nichols (*loc. cit.*) gives the following table:

|  | Number of Sons for 1,000 Daughters |
|---|---|
| Single births | 1,057 |
| Twin births | 1,043 |
| Triple births | 1,007 |
| Quadruple births | 548 |

Similar ratios are found in sheep.   The following table is attributed by Wentworth[1] to Pearl.   In 115 multiple sets in sheep the following sex-ratios occurred:

| | |
|---|---|
| 3 males to a birth | 16 |
| 2 males and 1 female to a birth | 39 |
| 2 females and 1 male to a birth | 22 |
| 3 females to a birth | 38 |

A summary shows that the ratio of females to males in these large plural births is 197 females to 148 males.

The reason for the preponderance of females in the plural offspring of these two species is not fully known; the basis of it is probably a sex-difference in prenatal mortality.   It is well known that the prenatal mortality of human male twins is greater than that of females, an indication that males are less resistant to abnormal uterine conditions than are females.   In the uterus of a species adapted for uniparous gestation the crowding of several fetuses must inevitably introduce untoward conditions; if males are more susceptible to subnormal conditions than females it is highly probable that they succumb in larger numbers than the latter.

That the mere fact of multiple offspring carries with it no necessary disturbance of sex-ratios is seen

[1] E. N. Wentworth, *Science*, N. S., XXXIX (1914).

when the ratios of normally multiparous species are examined. Wentworth gives the following tables (I and II) for swine and for dogs:

### TABLE I

#### SEX-RATIOS IN SWINE BIRTHS

| No. of males per litter .... | .0 | 1 | 2 | 3 | 4 | 5 |
|---|---|---|---|---|---|---|
| Expectation (in no. of litters)................ | 3.4 | 12 8 | 24.2 | 33 4 | 34.7 | 28.5 |
| Actual no. of litters....... | 2 | 13 | 26 | 28 | 31 | 28 |

| No. of males per litter .... | 6 | 7 | 8 | 9 | 10 | 11 |
|---|---|---|---|---|---|---|
| Expectation (in no. of litters)................ | 19 | 10 4 | 4.6 | 1.7 | 0.48 | 0.11 |
| Actual no. of litters....... | 21 | 12 | 8 | 2 | 2 | 1 |

### TABLE II

#### SEX-RATIOS IN DOG LITTERS

| No. of male pups per litter | 0 | 1 | 2 | 3 | 4 | 5 | 6 |
|---|---|---|---|---|---|---|---|
| Expectation (in no. of litters)............... | 15.1 | 35.75 | 37 | 24.5 | 10.86 | 2.8 | .3 |
| Actual no. of litters........ | 14 | 36 | 39 | 22 | 11 | 4 | 0 |

In the total of 173 litters of swine there is really no significant departure from the normal distribution of the sexes.

Again no disturbance of the expected ratio occurs in a total of 126 litters of pups.

From these statistics it appears that the disturbances in sex-ratios of plural births are limited to those species that are normally uniparous. This may be due partly to differential prenatal mortality, favoring the survival of females. There is nothing about these ratios at all out of accord with the idea that in mammals sex is zygotically determined.

## SEX-DIFFERENTIATION IN MAMMALS

Although sex is zygotically determined in mammals, the differentiation of sex-characters depends on a secondary mechanism that is believed to be associated with an internal secretion of the gonads. It has been long known that castration or ovariotomy of young mammals prevents the development of adult. sexual characters and the individual remains a neuter, though zygotically either a male or a female. Steinach in a brilliant series of experiments with rats has shown that a transplantation of ovaries into a young castrated male markedly alters the sex-differentiation of the operated individual, making it take on many of the characters of the female; even milk glands become functional and a maternal instinct develops. Conversely, a transfer of the testicular tissue into an ovariotomized female tends to masculinize the animal, so that it becomes large like the male and exhibits the pugnacious character of the latter. Since only the glandular portion of the transplanted ovary or testis survives in these experiments there is no alternative but to attribute the reversal of sex-tendency to some secretion of these glands, and, for a lack of a better term, the active principle has been called *hormone*. The interstitial glandular tissue of the testis secretes, therefore, male-differentiating hormones and the equivalent tissue of the ovary secretes female-differentiating hormones. These hormones must be given off into the blood, for it affects all parts of the body.

Crucial as are the experiments in transplantation of gonads, they do not equal in subtlety and finish the experiment of Nature performed in the case of the freemartin. Here an individual zygotically determined to be a female twin may become more or less completely differentiated into a male by the very neat device of borrowing hormone-charged blood from its male co-twin.

If two male twins mutually transfuse blood, no alteration of the sex-bias occurs. The same is true in the case of two female twins. But wherever the twins are opposite-sexed, the female is the one to suffer sex-reversal. At first it was not clear to Lillie whether this result was due to a dominance of male-differentiating hormones over female, or was the result of a precocious development of male glands. Further studies favor the latter interpretation, and it now appears that the glandular portion of the testis differentiates before that of the ovary and that when the twins unite blood vessels in the chorion, the blood of the male passes into the system of the female and inhibits the development of the glandular portion of the ovary. In the absence of female-differentiating hormones in the zygotically determined female, the male hormones have full play and actually do cause the differentiation of male characters in the freemartin. The freemartin is then a paradox: it is a zygotic female, but differentiated more or less extensively into a male.

These results make it necessary then to distinguish most carefully between sex-determination and sex-differentiation. Sex may be determined zygotically, but may be altered more or less completely by a change in the hormones. The chromosome mechanism appears to fix the sex-bias and the hormone mechanism to bring about sex-differentiation. Normally, however, the zygotically determined sex remains unaltered, for a zygotic female will produce female-differentiating hormones and will produce an adult female individual.

## IS TWINNING HEREDITARY?

In the polyembryonic armadillos (*Dasypus novemcinctus* and *D. hybridus*) it goes without saying that their peculiar process of twinning is hereditary; it is the only mode of reproduction in these species. What is really inherited in these cases is not fully understood, but it is believed that its basis lies in some physiological peculiarity of the egg, which causes it to have an abnormally slow early development. As brought out in the earlier chapter, this retardation in the developmental rhythm produces an early fission of the embryonic materials into several distinct primordia each of which produces an embryo. Whatever the cause of polyembryony, it is unquestionably a specific character and therefore hereditary.

In the armadillo *Euphractus villosus*, and possibly in other twinning species, the production of dizygotic twins is practically a specific character and is therefore inherited. The inherited character is some physiological peculiarity of the ovarian rhythm resulting in the synchronous ovulation of an egg from each ovary. In close relatives of this species the ovaries alternate in functioning, so that the right ovary would function in one pregnancy and the left in the next. The breaking up of this alternating rhythm in *Euphractus* and the establishment of a synchronic rhythm in the two ovaries would appear to involve a profound alteration of the general metabolism. That this altered condition is established as an inherited specific character is the only conclusion that the facts admit.

The question of whether twinning in ruminants is hereditary is dealt with by various writers. The experiments of Alexander Graham Bell furnish the best available evidence on this point. In 1889 he purchased a farm at Beinn Bhreagh in Nova Scotia and found himself in possession of a flock of sheep. In the spring of 1890 one-half of the lambs born on the place turned out to be twins. Evidently he had a race of sheep with an unusually strong twinning tendency, and it became an object of interest to discover just which ewes were twin-bearing and whether they showed any easily recognizable correlated characters. Bell says

> Upon examining the milk-bags of the sheep, a peculiarity was observed which was thought might be significant. Normally, sheep have only two nipples upon the milk-bag, but in the case of several of the sheep examined, supernumerary nipples were discovered which were embryonic in character and not in a functional condition. Some had three nipples in all, and some four. Of the normally nippled ewes 24 per cent had twin lambs; but of the abnormally nippled 43 per cent had twins. The total number of ewes, however, was so small (only 51) as to deprive the percentage of much significance. Still the figures suggested a possible correlation between fertility and the presence of supernumerary nipples, and it seemed worth while to make an extended series of experiments to ascertain (1) whether, by selective breeding, the extra nipples could be developed so as to become functional, and (2) whether the ewes possessing four functional nipples instead of two would turn out to be more fertile than other sheep and have a larger proportion of twins.

Apparently no difficulty was experienced in developing the embryonic supernumerary nipples into functional organs, and by 1904 nearly all of the ewes on the farm possessed four functional nipples and a practically pure line of sheep with four milk-bearing nipples was established. Subsequently many five- and six-nippled sheep appeared and occasionally seven- and eight-nippled ones were produced. The result was due, not to the selection of fluctuating variations, but to the appearance of mutations and the selection of these suddenly appearing new types as the parents of succeeding generations.

The attempt to establish a correlation between supernumerary nipples and twin-bearing was disappointing. In a subsequent paper Bell says:

> At first it appeared that four-nippled ewes were *less* fertile than ordinary sheep, for they had a smaller proportion of twins; but this turned out to be due to the fact that the process of selection had necessarily resulted at first in a flock composed mainly of young ewes, and young sheep rarely have twins. After the four-nippled ewes had grown to full maturity they were found to be as fertile in this respect as ordinary sheep, if not more so.
>
> The four-nippled stock proved a failure in so far as twinning was concerned, so in 1909 the flock was cut down to six-nippled ewes alone. There were indications in 1912 that the six-nippled stock will ultimately turn out to be twin-bearers, as a rule, when they become fully mature.

Whether or not this expectation was realized I am not in a position to say. On the whole, it does not appear that the production of a twinning race of sheep by selecting for supernumerary nipples is a success, for at best only a very general correlation between the twin-bearing tendency and the tendency to extra nipples exists.

A much more promising method of producing a twin-bearing race of sheep would be to breed exclusively from twin individuals irrespective of, or in correlation with, the extra-nipple character. Bell had in mind such a procedure when he wrote his 1912 communication, but has not, so far as I am aware, carried it out. He noted, however, several interesting facts that might increase the hereditary tendency to produce twins. "Twin-bearing ewes are on the average much heavier than single-bearing ewes." The condition of the mother at time of mating is also important, for when the mother is fat and in prime physical condition the percentage of twins is larger than when the mother is lean. Ewes mated in October when the pasturage is at its best have a much larger proportion of twins than those mated later in the breeding season when the pasturage is on the wane. Very few ewes mated in December have twins. A lowered nutrition of the mother after mating in October favors carrying twins successfully to birth, as it keeps the size of fetuses rather small and thus obviates undue crowding.

We may conclude from Bell's experiments that twinning is distinctly a hereditary character in sheep which is not merely sporadic but more or less racial; a fairly large percentage of twins appears to be specific for sheep, but this percentage may be greatly enhanced by selective breeding from twinning strains. The economic importance of establishing a twinning race of sheep is obvious, for, as Bell says, "if the farmers could raise two lambs instead of one for every ewe wintered, sheep breeding in Nova Scotia might become a profitable industry of great importance."

Although there is a widespread belief among breeders that twinning in cattle is hereditary, there is, so far as I am aware, no direct evidence that such is the case.

No experiments of the sort carried out upon sheep by Bell have been made upon cattle. Doubtless similar results could be obtained, although the normal percentage of twins in cattle is very much smaller than in sheep. Even if the ratio of twin births in cattle could be increased by selective breeding, the advantages of such an increase would be considerably reduced by the frequency of sterile freemartins.

There are, however, certain rather indirect evidences that dizygotic twinning is hereditary in cattle. Several cases of cows producing several sets of twins have been recorded by various writers. One interesting case, cited by Pearl, is that of a cow that produced successively three single offspring, then two pairs of twins, next triplets, then a single calf, and finally the set of triplets shown in the photograph (Fig. 40). The middle calf is a normal male and the two outside ones are freemartins. The male is a typical Guernsey like the dam, but the freemartins are therefore like their sire. The two freemartins are remarkably alike and are believed by Pearl to be monozygotic.[1] Cases of triplets are extremely rare, but Cole records seven cases among 303 plural births. Since plural births in cattle occur in only about 2 per cent of births, triplets occur only about once in 2,000 cases. Possibly many more triplet gestations begin, but result in the death or early abortion of one or more members of the set.

Since no experiments have ever been performed by way of selecting for a twinning strain of human beings, the only evidence of a hereditary tendency in twinning

[1] In another place I have shown the improbability of monozygotic twinning in cattle.

is statistical. Danforth[1] in a popular article concerning the heredity of twinning has collected data that may readily be interpreted as indicating that the tendency is not merely sporadic, but has a congenital basis. Fifty pairs of newborn twins were found to have 171 singly born and 10 twin older brothers and sisters, a ratio of 1:18. In mothers' fraternities (i.e., brothers and sisters) there were 318 single births and ten pairs of twins (1:32), and in fathers' fraternities 219 single and eight pairs of twins (1:37). When these ratios are compared with the normal incidence of twins (1:90.6), it appears that certain strains are more liable to twins than others; this implies a hereditary tendency.

I am well acquainted with a family in which strikingly similar duplicate twins occurred in two consecutive births. The first pair were females, the second pair males. A collection of data of this sort would be very interesting, as it would tend to indicate that monozygotic twinning is hereditary; no other such data are available to me.

Whether the tendency to twinning is a factor resident in the mother or the father is not clear. That the twinning factor, whatever it may be, might reside in the father is suggested by an extraordinary case cited by R. Berger.[2] The case concerns a man whose first wife had quadruplets once and twins ten times; his second wife had triplets three times and twins ten times. The man was the father of sixty-eight children. Dr. Berger inferred that the tendency to twinning is

---

[1] C. H. Danforth, *Journal of Heredity*, VII (1916)

[2] *Zentralblatt f. Gynäkologie*, X (1914).

due rather to the father than to the two mothers. How such a condition could be determined by the male is difficult to imagine, but it may well be that the tendency to polyembryonic development of an egg might be stimulated by some peculiarity of the sperm. If the twins in this case were all duplicates, this would be an interesting possibility; but the father could hardly control double or triple ovulation in the mother. Unfortunately no data are given as to the sex of the twins in this striking case.

It will readily be seen that monozygotic and dizygotic twinning involve totally different situations and would therefore doubtless be inherited in quite different ways, if inherited at all. Danforth is inclined to believe that the ability to produce twins is common to all strains, and that there is merely a variation in frequency of incidence of twins in different strains. The only method of settling the question is that of collecting a really adequate mass of data; no such collection is yet at hand.

Both monozygotic and dizygotic (including polyzygotic) twinning are characters capable of being inherited as unit characters and of being made racial or specific by selective breeding. In cattle and probably also in man twinning seems to be recessive and single births dominant; hence a pure twinning strain could be produced, if at all, only by interbreeding homozygous recessive individuals, that is, males and females that are the offspring for at least two generations of ancestors that were themselves twins.

# CHAPTER VII

## VARIATION AND HEREDITY IN TWINS

### A. VARIATION AND HEREDITY IN ARMADILLO QUADRUPLETS

The known laws of heredity and theories as to the mechanism of inheritance are at present applicable only to those usual reproductive modes which involve the development of but a single offspring from a fertilized egg. A new situation appears when an egg gives rise to two or more offspring, and new modes of inheritance are involved.

If polyembryonic offspring were absolutely identical within a monozygotic set, the problem would be to account for the identity. Since they are not identical, but show a definite variability within a set, the problem is to account for the differences that exist among them.

Only in one character are the members of a polyembryonic set always identical: they are always of the same sex. In all other respects intra-set differences of a more or less radical character exist. Some of these differences are purely extrinsic in nature or origin; others are strictly intrinsic or due to factors inherited from the parents.

The following problems connected with heredity in monozygotic quadruplets present themselves for solution: (*a*) What kinds of character are inherited? (*b*) Which ones are inherited by all and which are subject to unequal distribution among the fetuses of different sets? (*c*) According to what laws are such characters inherited? In the briefest possible way an attempt will be made to present an outline of some of the studies on heredity in armadillos which have been published in several recent papers.[1]

### MATERIALS FOR THE STUDY OF HEREDITY

No other animal is so beautifully adapted as is the armadillo for the detailed and accurate comparison of parent and offspring or of offspring among themselves. The strikingly diagrammatic arrangement of the integument into five armor shields (see frontispiece), each shield consisting of well-defined units (scutes or scales), furnishes an unparalleled opportunity for the study of inter- and intra-individual correlation. These characters, moreover, are definitive in number and arrangement long before birth. How fortunate a circumstance that this species, which has so unique a method of reproduction, should also possess equally unique possibilities for biometric treatment!

Although any part of the armor would serve well the purposes in hand, the banded region, on account of its regularity and clean-cut character, seems almost made to order for biometric study. Each band is made up of from 50 to 70 units, here called *scutes*. A scute consists of a horny scale, a bony base, and a definitely arranged group of hairs. For our purposes this whole complex may be viewed as a unit character. Studies have been made of the specific variability in total numbers of scutes in the nine bands and of that

[1] H. H. Newman, *Biological Bulletin*, XXIX, Nos. 1 and 2; *ibid.*, *Journal of Experimental Zoölogy*, XV, No. 2.

for each band. These studies are a necessary preliminary for determinations of the coefficients of correlation among quadruplet sets, but need not be dealt with
here. To illustrate the ways in which heredity works
in polyembryonic species only the most essential facts
about the modes of inheritance of these scute groups
need be presented.

In all of this work a limitation upon any complete
analysis of the situation is imposed by the fact that it
has been possible to study the heredity from one parent
only. Breeding in captivity was not found feasible
for several reasons. First, so large a number of sets
was required for statistical study that it would have
been an enormous undertaking to capture and keep the
necessary number of parents. Secondly, attempts to
keep armadillos in confinement showed that, as a rule,
they become badly diseased and die. Thirdly, in order
to obtain knowledge of the pairing and symmetry relations of the embryos, it was found necessary to remove
the unborn fetuses from their mothers; this involved
killing the mothers, a practice hardly feasible in the case
of painstakingly domesticated animals. Finally, in the
few cases where offspring were born in captivity, it was
found that the mothers ate the offspring, thus totally
nullifying the results of breeding experiments. Consequently our method of capturing and killing pregnant females, removing and preserving the fetuses, and
also preserving the armature of the mothers for comparison with those of the fetuses, gave the maximum results
possible with the material.

This limitation of the study of heredity to the
maternal side only is less of a disadvantage than might

at first appear, for the armor characters, which are the most available features for study, are the same in both sexes. There being no sex-dimorphism on the basis of these characters, the inheritance from mothers is equally strong for male and for female offspring. Consequently we have every reason to believe that what we discover about the inheritance from the maternal side would prove to be the same as that from the paternal side if the latter were known.

*Inheritance of the numbers of scutes.*[1]—The first study of inheritance of scute numbers was based upon a comparison of the total number of scutes in the nine bands in mother and in quadruplet offspring. A few type cases may first be cited:

Type I. *Maternal number dominant:* Set C 4. Mother has 574 scutes; fetus I, 574; fetus II, 579; fetus III, 571; fetus IV, 576.—Obviously all four fetuses are extremely close to the mother in this character. The resemblance is exact in the great majority of individual bands.

Type II. *Paternal number dominant in all four fetuses:* Set K 73. Mother, 541; fetus I, 521; fetus II, 518; fetus III, 523; fetus IV, 520.—Obviously none of the fetuses have inherited scute numbers from the mother. Presumably they have inherited it from the father, which had probably about 520 scutes.

Type III. *Maternal number present in some of the fetuses but not in others.* Three subtypes may be cited:

1. Set K 54. Mother, 565; fetus I, 565; fetus II, 576; fetus III, 569; fetus IV, 568.—This set shows three like the mother and one, presumably, more like the father.

---

[1] A study of the variability in numbers of scutes in the banded region for a large sample of the species shows that a range of from 517 to 625 scutes and an average deviation from the mean of 15 scutes. Compare with the low variability within the monozygotic sets.

2. Set K 13.  Mother, 565; fetus I, 575; fetus II, 570; fetus III, 563; fetus IV, 563.—Here we have one pair like the mother; the other pair, presumably, like the father.

3. Set C 29.  Mother, 561; fetus I, 549; fetus II, 560; fetus III, 547; fetus IV, 546.—Here we have one like the mother and three, presumably, like the father.  In this connection it is interesting to recall the relations of fetuses shown in Figs. 21 and 22, where it appears that three fetuses may be derived from one half of the ectodermic vesicle and one from the other; we would expect the three from one side to be much alike and the one from the other side to be somewhat different.

Type IV. *None of the fetuses are at all closely like the mother and still show wide differences among themselves:* Set C 65.  Mother, 543; fetus I, 561; fetus II, 563; fetus III, 573; fetus IV, 573.—The paternal number is probably at least as high as 573; the others are partly paternal and partly maternal.  Examination shows that the first 5 bands are much like those of the mother in fetuses I and II, while the last 4 bands are much like those of fetuses III and IV, which are probably purely paternal.

It is not rare to find good cases of types I and II which show nearly pure maternal or paternal dominance of all nine bands.  Nearly 25 per cent of cases could be classed in each group.  The remaining cases fall in the classes that show the various inter- and intra-individual distributions of maternal and paternal dominance.  One comes to the conclusion that the armor characters are examples of mosaic or particulate inheritance.  The maternal influence is largely dominant in some sets, or in some individuals of a set, and the paternal in others.  There are also cases in which the influence of one parent predominates in some bands, and that of the other parent in other bands.  Finally, there are some cases in which one lateral half of the body has quite a different number of scutes from the other half, and one of these halves resembles the maternal condition. For tables showing all the data upon which this discussion is based the reader is referred to a special paper on the inheritance of numbers of scutes.[1] There a study of inheritance was made for each of the five armor regions. What has been said for the banded region applies equally well to the other four shields.

A general conclusion from this intricate and extensive mass of statistical data is that both large and small groups of integral variates, such as the aggregate of scutes in an armor shield or a single band, are inherited primarily according to the Mendelian laws of dominance, with only a minor degree of blending, and that the dominance is regional and not very often general for a large section of armor. This, then, is a sort of particulate inheritance, a mode of inheritance which *implies a somatic segregation of parental characters.*

When biometrical methods of inheritance study are applied to this material, certain new facts are brought out, but certain other facts already seen for individual cases are reduced to general terms and are likely to be lost sight of.

*Correlation between mothers and offspring as to numbers of scutes.*—By the use of standard statistical methods the coefficients of correlation between mothers and offspring have been derived for a large number of characters. To illustrate: it was found that, with respect to the total number of scutes in the banded region of armor, there was a coefficient of correlation between mothers and 56 sets of male quadruplets of 0.5522±0.0625, and for 59 sets of female quadruplets,

[1] H. H. Newman, *loc. cit.*

0.5638±0.0597. Making allowance for probability of error, the coefficients of both sexes are practically identical; the coefficient is about 0.5, which is just what we should expect if mother and father contributed equally to the inheritance of these characters. Presumably, then, the coefficient for fathers and offspring would also be 0.5 or thereabout. It will be noted also that the males inherit from mothers just as strongly as do the females, which goes to show that we can ignore sex in dealing with the inheritance of scute characters. Practically the same degree of correlation exists for the number of scutes in the other four parts of the armor. All of this evidence simply means that, on the average, embryos show as much paternal dominance in scute numbers as they do maternal. Remembering, then, that the degree of resemblance between mothers and their polyembryonic sets of offspring is about 0.5, it will be interesting to compare with this the degree of resemblance among the quadruplets themselves.

*Correlations among individuals of quadruplet sets as to numbers of scutes.*—Using the same statistical methods as in determining the heredity between mother and offspring, we find that for total numbers of scutes in the banded region there is a coefficient of correlation for 56 sets of male quadruplets of 0.9294±0.0057, and for 59 sets of female quadruplets, 0.9129±0.0059. This degree of correlation is extremely high and it has no parallel among interindividual correlation coefficients. In fact, the members of quadruplet sets are as strikingly alike (as closely correlated) as are the paired organs of single individuals or as the right half of a single individual is like the left. The correlation constants brought out for the numbers of scutes in the banded region are practically duplicated by those derived from a study of any other part of the armor. Results of this sort would in themselves be sufficient to prove the polyembryonic character of armadillo quadruplets; if these individuals were from four germ-cells there would be no reason to expect correlations higher than those that obtain between brothers, which are never much higher than 0.5. From the genetic standpoint a set of armadillo quadruplets is essentially one individual in four parts. When we are comparing one fetus of a set with another, we are dealing with *a special case of intra-individual correlation*. The proof of this assertion lies in the fact that the degree of correlation is of the same order as those determined for equivalent parts of one individual and is never paralleled by that determined between separate individuals.

In closing this very brief statement of some of the significant facts about the inheritance of aggregates of integral variates, two points should be emphasized. First, it should be said that these scute number differences are about the only inherited differences found in the species that are available for biometric study. Armadillos are practically all alike in color, size for age, and proportions of body parts. Males and females are also alike except for the genitalia. In view of these circumstances the reader will readily appreciate my choice of characters for the study of heredity. Whatever facts about heredity are to be discovered for this species must, therefore, be discovered in connection with these scutes of the armor. The second point to remember concerning the armor characters is that they are inherited in mosaic fashion.    There is every evidence of an unequal distribution of inherited units among the cleavage products of the single germ-cell.    This results in an interindividual segregation of inherited characters, which is much like the unilateral appearance of a color character in certain piebald types, where one half of the face is colored, the other half white, or where one foreleg is colored and the other not.    This somatic segregation of inherited characters is very general and will be discussed more at length in a subsequent connection.

*Inheritance of double bands.*—The arrangement of the scutes of the banded region is in general remarkably regular (see frontispiece).    Each band is typically composed of a single row of scutes.    A small percentage of individuals show an irregularity in scute arrangement consisting of parts of bands that are double, while the rest are single.    In Fig. 42 a number of types of band doubling found in a single set of quadruplets are shown. Sometimes the double part is quite extensive, involving a large part of the band; sometimes it consists of a doubling of only one scute (third band from top to right, Fig. 42).    All intermediate conditions are found. Band doublings may be confined to one half-band, or they may be repeated on both halves; i.e., they may be bilateral or unilateral in their expression.    The presence of irregularities of this sort furnishes us with the only data by means of which we can get a really definite idea of the distribution of inherited characters among polyembryonic offspring.    It was noted quite early in comparing the individuals of polyembryonic sets that sometimes band doublings were repeated with striking faithfulness of position and detail in two or more individuals of a set and were totally absent in others of the same set. Sometimes all four individuals showed these characters, but to a very different extent or in different positions. For example, the doubling might be unilateral in one pair of twins and bilateral in the other, or the character might involve a dozen scutes in some and only one or two in others. The real significance of this situation did not become apparent until it was found that these somewhat anomalous arrangements of scutes, which in general we call "doublings," were definitely inherited. It appears that there is a close genetic relation between "scute doublings," where the anomaly affects only one armor unit (an incipient doubling), and "band doubling," where from two to many units are involved. Sometimes band doubling in the mother is inherited in the offspring as scute doubling, and vice versa. Quite often the expression of the character may differ within the set of offspring, so that a band doubling or a scute doubling in the mother may be inherited in some offspring as a band doubling and in others as a scute doubling.

In general it may be stated that in all cases except two, which are quite doubtful, when a mother has either a scute or a band doubling, one or more offspring in a set show a doubling. There are three categories in one homogeneous collection of 140 sets of quadruplets, in which the condition of the mother is definitely known:

(a) Those in which both mother and offspring show doubling; of these there are 56 sets, 29 female and 27 male.

(*b*) Those in which the mother is normal (without doubling), but in which doubling occurs among the offspring; of these there are 41 sets, of which 22 are female and 19 male; in the group it is assumed that the father possessed the factors for doubling.

(*c*) Those in which both mother and offspring are entirely without doubling; of these there are 43 sets, of which 22 are female and 21 male; in this group it must be assumed that the father possessed no doubling factors.

These data would appear to show conclusively that the "doubling" factor is inherited as a dominant, but that the mode of inheritance is not typically Mendelian; if "doubling" is dominant and "lack of doubling" is recessive, we should expect a considerable number of individuals to be heterozygous for "doubling" and to produce equal numbers of germ-cells that carry the doubling factor and of those that do not. If heterozygosity occurred, we should often get "non-doubling" in sets of offspring from mothers that show doubling but are heterozygous for the character. That this result is not realized is a strange circumstance, and one that can be explained satisfactorily only on the assumption that a segregation of dominant and recessive factors occurs during cleavage so that the blastomeres, which go to produce both soma and germ-cells of the individuals, are pure for the factor in question, and that homozygous offspring are always produced, and never any heterozygous ones. Segregation like that which is supposed to occur during maturation divisions, when chromosome reduction accompanies it, would appear to occur here during the process of cleavage in which no reduction of chromosomes takes place. But before entering upon a discussion of somati⌐ segregation, we must needs study some of the data upon which the theory is based. Only a few selected cases may be presented within the scope of the present volume; the reader is referred for complete data to two recent papers.[1]

· For convenience of presentation it is necessary to conventionalize the figures of band and scute doubling. The methods of representing band doublings together with the detail of actual cases are shown in Fig. 42. At the top of the page is shown a single band with an extensive doubling involving all but a few marginal units at the right and the left. Directly underneath is a conventional representation with numbers indicating the numbers of scutes involved. Below this is another type of doubling in detail, with the conventional representation just beneath. Various types of scute doubling are shown also, and the method of indicating the location and distribution of them is by placing a small diagram of a double scute in a band and locating its position with reference to margin or middle by a number. In doubtful cases an arrow points to the spot from which the count proceeds.

The character and distribution of inherited band and scute doubling may be illustrated by a complete detail drawing of the affected bands in one set of fetuses (set K 87) and their mother (Fig. 42). The affected band of the mother is marked *M* and those of the four fetuses I, II, III, and IV. Fetuses I and II (a pair) have each two affected bands. It will be seen that the mother has a unilateral doubling involving 13–14 scutes six places from the left-hand margin of band 1.

[1] H. H. Newman, *loc. cit.*

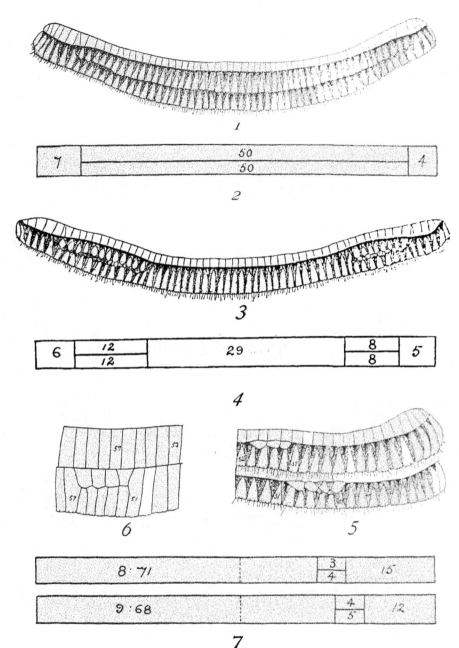

FIG. 41.—Various kinds of band doublings, and conventional methods of representing them. The detail of a double band is shown in *1* and a conventionalized diagram of the same band in *2*. Another kind of band doubling is shown in detail in *3*, and the same in diagram in *4*. Similarly, *5* is a detail of scutes and *6* a detail of underlying bony plates. In *7* are diagrams of two adjacent bands with similar doubling in both; 8:71 and 9:68 mean band *8* with a total of 71 scales and band *9* with 68 scales. The other numbers indicate the number of scutes in the various regions.

All four fetuses have a more or less extensive doubling of the same band. In fetuses I, II, and IV the doubling is bilateral, but the more extensive doubling is on the

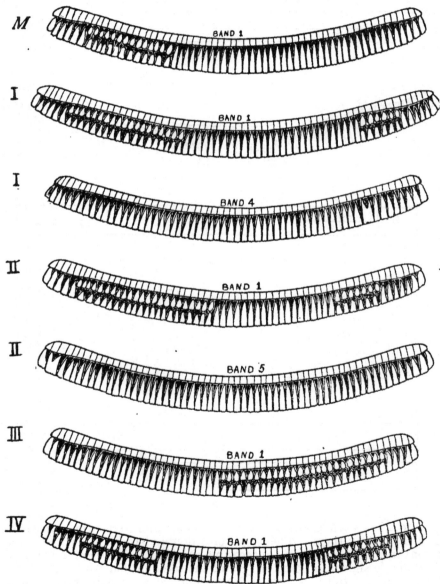

FIG. 42.—A detail drawing of all the bands that show doubling in mother and offspring of set K 87. *M* = mother; I, II, III, IV, the four fetuses.

left side; in fetus III it is unilateral but on the side opposite to that of the mother. Fetuses I and II have in addition, in bands 4 and 5, respectively, a double scute each. In fetus I the double scute is near the right margin; in fetus II it is near the left margin. This type of symmetry reversal is very common between individuals of a pair and is called mirror-imaging. Fetus I (belonging to one pair) has the more extensive doubling on the left side, and the opposite fetus, III, (belonging to another pair) has all the doubling on the right side; this is an example of mirror-imaging involving embryos derived from opposite sides of the egg. That this is a genuine case of inheritance of doubling can scarcely be doubted when it is considered that band doubling occurs in only about 3 per cent of individuals in the species. Its occurrence in mother and four off-spring is more than a coincidence.

Set K 30 (female fetuses) further illustrates the mode of inheritance of scute and band doubling. A somewhat simplified method of representing the anomalies is here adopted (Fig. 43).

In the mother there is in band 1 a minimal band doubling involving two scutes located six places from the left margin. In band 2 there is a scute doubling ten places to the left of the middle. Some kind of doubling appears in band 1 of all four fetuses. Fetus I has a double scute in exactly the same spot where the mother has an incipient double band; fetus II has in the exact middle of the band a short band doubling of three scutes and a double scute 14 places to the left of the middle, or nearly in the place where the mother has a double scute in band 2. In addition, fetus II

has a double scute in band 5 quite close to the middle
or almost directly in a line with the band doubling

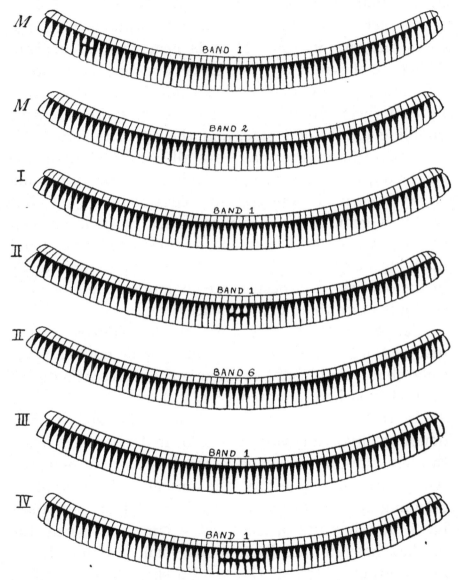

FIG. 43.—A detail drawing of band doubling in set K 30

farther forward. Fetus III has a double scute exactly
in the middle of band 1; and its partner, fetus IV, has

a band doubling involving seven scutes exactly in the middle of band 1. Nothing could show more clearly the genetic equivalence of scute and band doubling, for obviously the three doublings exactly in the middle of band 1 in three fetuses (II, III, IV), one involving one, another three, and another seven scutes, are genetically equivalent and must be inherited from the condition in the mother. It is quite common to find positional reversals involving a shift from near the margin to the middle. This type of symmetry reversal is scarcely mirror-imaging, but is nevertheless of a kindred character, doubtless due to similar factors. It will be noted that fetus I has the same position of the doubling as the mother has in the incipient band doubling, while fetus II has inherited the double scute of band 2 of the mother in its band 1. Fetuses III and IV (a natural pair) have a positional reversal of that of the mother, III inheriting a scute doubling in the same place where fetus IV inherits a band doubling. Again, it will be seen that only the primary fetuses II and IV inherit the band doubling, while the two secondary individuals inherit the scute doubling. Fetus II is the only one that inherits both scute and band doubling and has the doubling involving two bands, as in the mother.

Set K 4 (male fetuses) shows another set of conditions (Fig. 44). The mother has in band 1, beginning seven places from the left margin, a small band doubling of three scutes. Fetus I has in band 1, four places to the left of the middle, a somewhat more extensive band doubling involving five scutes. This is evidently a case of inheritance with symmetry reversal of one side. All the other fetuses have extensive band doubling. In

fetus III the doubling involves the whole band, for
there is an extra or tenth band present; in fetus I the
whole band is double except four scutes on the left

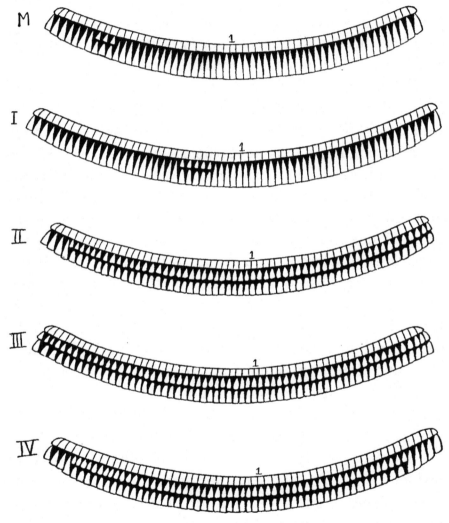

Fig. 44.—A detail drawing of band and scute doubling in set K 4

margin; and in fetus IV the whole is double except four
scutes on the left and five on the right. Fetuses I
and II (a pair) are alike in being unilateral in doubling,

and III and IV (a pair) are both bilateral. Perhaps the most striking feature illustrated by this set is the great variation in the extent of doubling that may appear in the four fetuses of a single set where there can be no question as to the common genetic basis of the more and of the less extensive expression of the character.

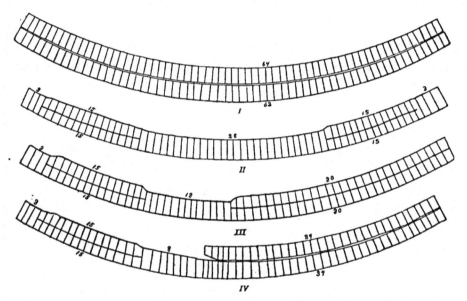

FIG. 45.—Diagram of band doubling in set A 64

By way of contrast, however, we shall present a case in which the mother showed no doubling, but in which the fetuses show a remarkable degree of resemblance in the position and extent of the anomaly. Presumably the condition has been inherited from the father. Fig. 45 is a diagrammatic representation of the doubling of bands in Set A 64. Fetus I has the entire first band double, for there are ten bands. Fetus II has a beautifully symmetric bilateral doubling of band 1, consisting of fifteen double scutes situated three places from the right and from the left margin. It will be noted that, while both fetuses of this pair are bilaterally symmetrical in the anomaly, fetus I has complete doubling and fetus II has incomplete doubling. Fetuses III and IV have absolutely identical bilateral doubling, but are asymmetrical in that the left side of each has the type of doubling in every detail of II, while the right side of each is completely double like the condition in fetus I.

These cases of band doubling must serve as samples in that they show practically all the peculiarities involved

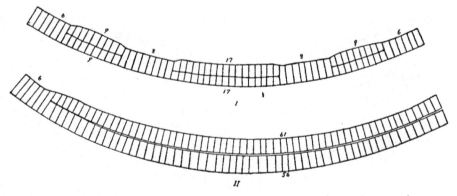

FIG. 46.—Diagrammatic representation of doublings in set A 101

in band doubling except those in which only one, two, or three of the fetuses in the set show a doubling. One case of that sort must be cited.

Set A 101 (male fetuses) is a case in which one pair has band doubling and the other has no doubling at all (Fig. 46). Fetus I has an exact bilaterally symmetrical anomaly of band 1, involving three double regions, two lateral ones of nine scutes each, six scutes from the margins, and a median doubling of seventeen scutes. Fetus II has the band double except six scutes on the left margin. We must consequently inquire why doubling, evidently inherited from the father, since the mother showed none, could be confined to the two fetuses and excluded from the other two, when all came from the same fertilized egg.

*Data on the inheritance of scute doubling.*—Scute doubling is far more frequent in its incidence than band doubling, but the modes of inheritance and distribution are the same. Band doubling may be inherited as scute doubling, or vice versa. A few examples of the inheritance of scute doubling will be sufficient to illustrate the principles involved. Impressive cases of resemblance between mother and offspring and among the quadruplets of a set are of frequent occurrence.

Set K 27 (Fig. 47) is one of the most remarkable cases of exact inheritance. The mother has two double scutes located in definite places in two different bands. One of the offspring (fetus I) has two identical double scutes in exactly the same places in the same two bands as in the mother. The other three fetuses are without any doubling.

Set C 26 (Fig. 48). The mother has a double scute of a peculiar kind two places from the left margin of band 2. The paired fetuses (III and IV) each have identical double scutes in exactly the same position in band 1.

Many other cases occur of close resemblance between mother and offspring involving, however, a reversed symmetry. An example will illustrate this:

Set C 76 (Fig. 49). The mother has in band 6 a double scute five places from the *left* margin. Fetus III has a similar double scute in band 4 five places from the *right* margin.

Set C 30 (Fig. 50).  The mother has in band 1 a double scute two places from the *right* margin, and fetuses II, III, and IV have each a double scute two places from the *left* margin of the same band.

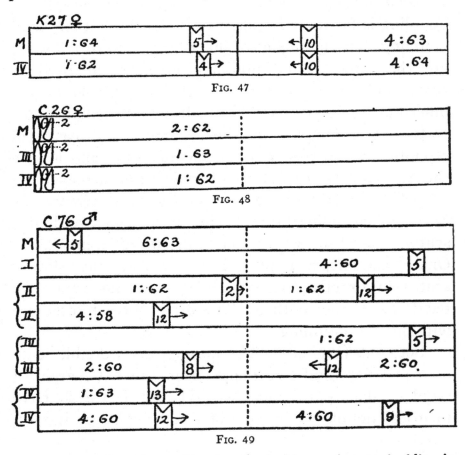

FIG. 47

FIG. 48

FIG. 49

FIGS. 47, 48, and 49.—Showing the incidence of scute doubling in sets K 27, C 26, and C 76, respectively.  (Details in text.)

*Mirror-imaging among the fetuses of a set.*—It is very common to find symmetry reversals within a set of quadruplets involving what we call mirror-imaging.

Set C 30 (Fig. 50).  Fetus I has a peculiar type of scute three places from the *right* margin of band 1,

while its partner has an identical double scute two
places from the *left* of band 1. Fetuses III and VI
(the other pair) have the same element distributed

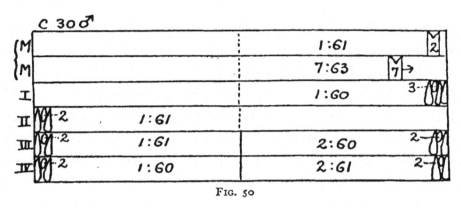

FIG. 50

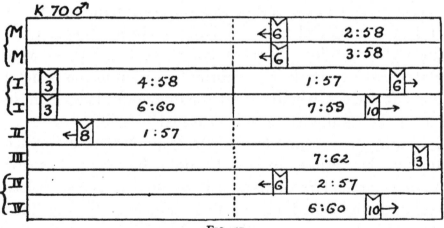

FIG. 51

FIGS. 50, 51.—Showing the incidence of scute doubling in sets C 30
and K 70.  (Details in text.)

bilaterally but in band 1 on the left and in band 2 on
the right.

Set K 70 (Fig. 51).  Fetus I has a double scute
three places from the *left* margin of band 6, while fetus
III, belonging to the opposite pair, has a similar element

3 places to the *right* of band 7.   Note that these individuals are not members of a pair, but are situated back to
back on opposite sides of the vesicle.

*Half-band reversals.*—When in one fetus a double
scute occurs on, say, the right side near the margin and
in another fetus of the same set a similar element occurs
on the same half-band, but near the middle, there is said
to be a half-band reversal.   Such conditions are exactly
like the occasional reversals of pattern of the index fingers
of duplicate human twins (see Fig. 54), where the right
hands of the two show a reversal or mirror-image symmetry in one finger.

Set C 76 (see Fig. 49).   Fetus IV has a double scute
twelve places to the left of the middle of band 4.   Fetus
II has a similar element in band 1 twelve places from
the right margin.   Note that if these two fetuses were
placed back to back (their normal position in the vesicle)
the left halves would be mirror-image duplicates in so
far as the double scute is concerned.

Many more examples of extremely close resemblance
among the fetuses of a set might be cited, but for
further particulars the reader is referred to the complete
accounts published elsewhere.   Suffice it to say that
the number of cases in which the double scute is unequiv-
ocally inherited is so large that, even when the location
of the element in mother and offspring and among the
fetuses of a set is quite diverse, we must still conclude
that the character is inherited, but lacks the localization
factor.

This leads to a brief discussion of the possible factorial
basis of scute- and band-doubling inheritance.   There
appear to be at least four unit factors involved: (1) a

general factor for doubling; (2) several minor factors determining in double scutes the particular kind of doubling; (3) a factor determining extent of doubling, whether one element or many are involved in doubling; (4) a localizing factor, determining more or less precisely the exact position of the doubling in the individual. It may be supposed by way of illustration that a mother has the doubling factor, together with the extension factor; the father lacks the doubling and extension factor, but has the localizing factor. The factors could be variously distributed among the four fetuses so as to give all sorts of combinations. It seems quite evident, then, that doubling is a complex matter, not at all a simple unit factor, and it must therefore be explained on the basis of the interaction of several factors.

*Somatic and germinal segregation.*—That band and scute doubling are definitely heritable is proved by the fact that doubling is always present in some form in the offspring of mothers that show doubling. The real problem is to find a mechanism to explain why it so often happens that some of the fetuses derived from a single egg exhibit the character and others do not. In monozygotic quadruplets we should expect the same genetic constitution in each individual unless there exists some segregative mechanism, resulting during early ontogeny in an irregular distribution of the factors responsible for doubling. It has been suggested that the real differentiating factors are environmental; but the only environmental differences conceivable for monochorial quadruplets are nutritional differences due to more or less extensive placentation. It can be shown, however, that nutritional differences, so great as to have a pronounced effect on size and stage of development of different individuals of a set, have no effect on the inheritance of doubling. There are several quadruplets in which the various individuals are pronouncedly different in size, but are practically identical in the character and incidence of doubling. The differentiating factor, therefore, must be within the embryo itself. It seems logical to look to the cleavage mechanism as the probable seat of the irregular distribution to different areas of the blastoderm of the factors of doubling. Presumably, with an ideally accurate cleavage mechanism there would result an exactly identical incidence of doubling in all four fetuses of a given set. *That identity in doubling is not realized argues strongly for unequal distribution (or somatic segregation) of factors during cleavage.* Moreover, since doubling is evidently as strongly inherited from father as from mother, it seems probable that nuclear elements are chiefly involved, for the cytoplasm of the sperm cell is so small in amount as to be negligible.

If we assume that the beginning at least of segregation occurs in the first and second cleavages, we may suppose that when the factor goes to the first two blastomeres it is pretty certain to appear in both pairs of fetuses; when it goes to each blastomere of the four-cell stage it is likely to appear in all four fetuses. If, however, the distribution of the factor is such that it goes entirely to one of the first two blastomeres and not to the other, we should expect doubling to appear in only half of the fetuses; if, again, only one of the four cells gets the factor, we should have the factor in only one quadrant of the blastocyst and hence should probably find doubling in only one fetus.   This final resort to the early cleavages as the probable mechanism responsible for the distribution of doubling factors in polyembryonic sets is taken advisedly after a thorough canvass of all other possibilities which have suggested themselves.   The idea is not incompatible with the observations that have led to the budding hypotheses of Patterson, as has already been shown, but fits in better with the fission theory of polyembryony.

Segregation of unit factors during cleavage may be termed somatic segregation whether or not polyembryony be involved.   Somatic segregation must, I believe, take place in those forms that exhibit mosaic or particulate inheritance; in a spotted animal, showing in some areas the paternal and in others the maternal color, we certainly have a simple case of somatic segregation.

It is in connection with bud variations or clonal variations, however, that we find a condition more nearly analogous to what appears to happen in armadillo embryos. A green plant with colored flowers may produce as a bud variation a white-leaved branch with white flowers. If such a white flower were self-fertilizing and produced good seed from which white plants could be reared, we should have more than mere somatic segregation, since germinal segregation has gone hand in hand with somatic. I have been informed that just such a situation is sometimes realized for plants.

Now in the armadillo the four fetuses are produced agamically as fission products; they therefore constitute a clone.   There is every opportunity for clonal variation, and when a clonal individual gets a factor that makes for doubling in the soma, it always has gametes pure for this character. Otherwise why does a mother with doubling always have offspring that show some form of doubling?

Parallel somatic and germinal segregation can be explained only on the assumption that the segregation mechanism operates at a period prior to the differentiation of germinal and somatic cells, and this must be during the early cleavage stages.

*Mirror-imaging and symmetry reversals as the result of polyembryony.*—So frequent is the occurrence of mirror-imaging among armadillo quadruplets that it must be conceived as causally related with polyembryonic development. The two phenomena are so closely related that it is my belief that the occurrence of symmetry reversal or mirror-imaging in twins or double monsters may safely be taken as a criterion of their monozygotic origin. Only, however, when there appears some asymmetric feature like unilateral doubling is mirror-imaging recognizable. If both sides of twin individuals are exact bilateral duplicates no symmetry reversals would be possible; hence we must depend upon the unilateral appearance of such features as doubling for signs of reversed symmetry relations.

All grades of mirror-imaging are found in armadillo quadruplets. There may be mirror-imaging between individuals of opposite pairs, but this is much less common than imaging between twin partners derived from one half of the egg.

A still more frequent type of mirror-imaging is seen between the antimeric halves of a single individual. These facts lead to the conclusion that the symmetry relations among quadruplets are the result of an intricate interplay of three grades of successively operating symmetry systems, the later tending to obliterate the effect of the earlier, but not always successfully. In general, mirror-imaging between opposites is evidence of a residuum of a primary bilateral symmetry that held sway in the blastocyst before polyembryonic fission began. When the primary outgrowths are formed, they are the product of the antimeric halves of the first embryo and should therefore show mirror-image relations. But a partial physiological isolation of the two halves permits a certain reorganization or regulation of new symmetry relations, which tends more or less completely to destroy the original symmetry, yet often leaving a trace of the latter. Similarly, when the secondary outgrowths arise between the primary ones a certain residuum of the primary symmetry may be carried over that frequently manifests itself in mirror-imaging between twins derived from one half of the original embryo. Finally, when each secondary outgrowth organizes its own bilateral symmetry, it tends to lose, partially at least, the earlier symmetry relations, and to establish its own mirror-imaging of right and left sides. In some cases traces of all three symmetry systems appear in a single set of fetuses, but it is common to find only two systems interacting.

*Mirror-imaging between individuals limited to integumentary structures.*—When examples of symmetry reversal were first discovered to occur so frequently in the armor characters of the armadillo, it seemed probable that similar reversals would be found in the visceral arrangements. One would expect to find occasionally the heart apex turned to the right instead of to the left, and the greater curve of the stomach turned to the right. An extensive examination, however, shows that no visceral reversals occur. In his book, *Problems of Genetics*, Bateson calls attention to the same situation in human duplicate twins and says: "If anyone could show how it is that neither of a pair of twins has transposition of the viscera the whole mystery of division would, I expect, be greatly illuminated."

From what we know of the process of polyembryonic fission in the armadillo, it would appear that the most likely solution of this problem lies in the fact that twinning is initiated and carried out in the ectoderm, and the endoderm becomes involved only passively and considerably later. What more natural, then, than to look for evidences of twinning—mirror-image reversals—only in ectodermal and closely associated structures of the integument? In human duplicate twins it is true also that reversals are confined to the friction ridges, which are quite homologous in origin with the armor characters of the armadillo.

If Bateson should chance to read this suggestion as to the reason why transposition of viscera does not occur in twins, I doubt whether he would admit that "the whole mystery of division is hereby greatly illuminated." It is an interesting fact, however, that the ectoderm, where the rate of metabolism is highest, should take the lead and carry out the fission process, thus imposing the results of fission on the tissues of lower rate of metabolism, the endoderm and the mesoderm. The ectoderm is dominant in early development and produces the central nervous system through which it exercises progressively greater dominance as development proceeds.

### B. VARIATION AND HEREDITY IN HUMAN TWINS

No studies of value have appeared dealing with variation and heredity in dizygotic or fraternal twins; consequently our attention must be directed solely to these phenomena in duplicate (presumably monozygotic) twins. Perhaps the strongest evidence in favor of the idea that certain human twins are monozygotic comes from a study of the variability or lack of variability between these twins. So pronounced is the lack of variability in some cases that such twins are called "identical." A treatment of human twins paralleling that of armadillo quadruplets will show many points in common.

#### THE DEGREE OF IDENTITY IN HUMAN DUPLICATE TWINS

The first serious attempt to determine the closeness of resemblance between twins was made by Galton. As early as 1875 he showed his appreciation of the value of such a determination; this comes out clearly in his paper entitled "The History of Twins, as a Criterion of the Relative Power of Nature and Nurture." Galton conceived the idea that the degree of resemblance between duplicate twins is an index of the strength of heredity as opposed to environment. In attempting a minute comparison between twins he found them so much alike that he had to resort to such details as the patterns the friction ridges in the palms and soles.

It remained for Wilder, however, actually to carry out this study which Galton formulated, and some discussion of his (Wilder's) conclusions appears in the last few pages of this chapter.

*Wilder's studies of friction-skin patterns in the palms and soles of twins.*—It is a familiar fact that the surest method of identifying human beings is the finger-print method now in use in all modern detective bureaus and prisons. The method is based on the wide range of diversity in the details of the friction-ridge patterns

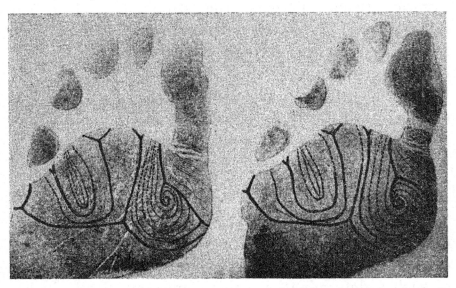

FIG. 52.—Photograph (from Wilder) of the left sole-prints of a pair of duplicate twins. The heavy lines are lines of interpretation. Note the striking similarity amounting almost to identity.

of the palmar surfaces of the hands and feet. No two individuals are exactly alike in all details, but the resemblances between certain types of twins are really surprisingly close. The prints of the left soles of a pair of twins studied by Wilder show identity of general pattern (Fig. 52) but lack of identity in detail, for the exact number of friction ridges in corresponding parts of the pattern differs in the two individuals.

Wilder has devised an elaborate method of classifying the various types of pattern in both palm and sole, and an equally elaborate system of condensed formulae for denoting the pattern complex of any individual. For a description of this method of formulating friction-skin diversity the reader is referred to Wilder's various papers on these subjects. It would lead us too far afield to attempt to introduce any adequate key to the study of human palms and soles in this place. Suffice it to say that in his studies of fifteen pairs of same-sexed twins and one set of opposite-sexed triplets he noted that eleven of them were "identical" in palm and sole patterns and five were unlike. The conclusion is reached that those which are "identical" are monozygotic and those which fall considerably short of identity are dizygotic. Since there is no knowledge of the intra-uterine conditions in any case, this reasoning backward from identity to origin is in this case unjustified. That this kind of reasoning may lead to grave error is shown in connection with those armadillo quadruplets in which there are quite pronounced differences between the members of a set; yet their origin from a single egg-cell cannot be questioned. In certain monozygotic human twins it might readily happen that by somatic segregation the paternal condition of friction ridges would go to one individual and the maternal to the other; or there might be a partial segregation of important elements of the pattern so that they would be dis tributed differently in the two.

It must not be lost sight of, however, that identity in friction-ridge patterns of twins makes their monozygotic origin very highly probable. Identity may

demonstrate monozygotic origin, but lack of identity
does not disprove the possibility of monozygotic origin.
Wilder has illustrated certain very good cases of identity
by means of photographs of palm- and sole-prints (see
Fig. 52). The heavy lines are the lines of interpretation

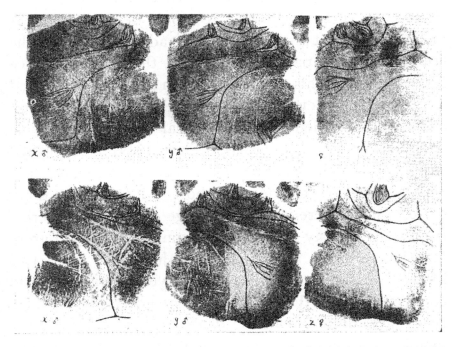

FIG. 53.—Photograph (from Wilder) of the left (above) and right
(below) palm-prints of a set of triplets. Note the close identity of the
males, which are evidently "identicals," and the unlikeness of these
to the female triplet on the right, which has evidently come from a
separate egg.

and serve merely to emphasize a real, fundamental
identity. Another very interesting case is that of triplets
(two boys and one girl, Fig. 53) and illustrates very well
the difference between ordinary "fraternal" resemblance
and true identity. The palm-prints of the girl are no
more like those of the two boys than is usually the case

for ordinary brothers and sisters, but those of the two boys are extraordinarily similar.   It may be decided that the two boys are monozygotic and that the girl was derived from a separate zygote.

The most significant features of the finger prints of duplicate twins are:

1. In one pair of twins there was an almost complete mirror-imaging of the two palm patterns of the two individuals.   The left hand of $x$ corresponds to the right hand of $y$, and vice versa.   This case corresponds to the rather rare mirror-imaging of "opposites" seen in armadillo quadruplets, and goes far to prove that polyembryony actually occurs in man.

2. In three other sets symmetry reversal occurs to a limited extent, but, curiously enough, always in connection with the pattern of the index fingers.   In two cases the finger-prints of the *left* index[1] finger of $x$ and of $y$ show a reversed symmetry so that one of them mirrors the condition in the left index finger of the other twin.   In the third case the symmetry reversal is in the *right* index finger.   These cases are quite homologous to the half-band reversals noted for armadillo quadruplets.   In human twins these reversals are fairly numerous considering the small number of twins examined.   Fig. 54 shows the finger-prints of one pair of duplicate twins with the reversed patterns in the two upper prints.

"These reversals of index patterns," says Wilder, "seem to occur with too great frequency to be disposed

[1] An interesting parallel exists between the conditions seen in man and in the armadillo.   In man mirror-imaging is usually confined to the index finger, while in the armadillo it is largely confined to the first band of armor.

of as a lack of correspondence with the rest." Yet no well-defined idea of the significance of these reversals seems to have occurred to Wilder. Bateson, however, sees in these peculiar phenomena evidence that twins derived from a single zygote have been parts of a single system of symmetry. This is evidently the key to the significance of symmetry reversals, as was brought out in the discussion of symmetry reversals of armadillo quadruplets.

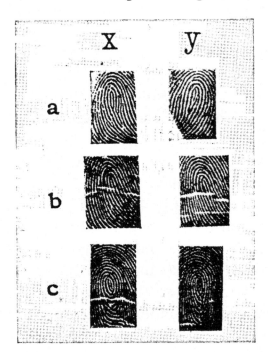

FIG. 54.—Prints (from Wilder) of the tips of the first three fingers of the left hand of a pair of "identical" twins. Note the reversed symmetry of the index-finger prints. This is a good case of mirror-imaging, so characteristic of monozygotic twins.

As Wilder has pointed out so clearly in his latest paper ("Palm and Sole Studies"[1]):

The bands of the armadillo carapace, with their variation and the friction ridges of human palms and soles, are partially or wholly homologous structures, so that their use in determining the degree of similarity of twinned individuals is equally warrantable in both cases, while the results may well be compared. Both deal with epidermic structures, the probable homologues of reptilian scales, placed in rows; in both are observed the similar phenomena of the forking and con-

[1] *Biological Bulletin*, XXX, Nos. 2 and 3 (1916).

sequent doubling of the lines thus formed. Even the double scale in the armadillo may have its counterpart in a twin sweat-pore, which indicates the composite nature of the unit to which they belong.

Wilder believes, however, that the far greater complexity of pattern in the human friction ridges gives a better basis for detailed comparison than the simpler condition seen in the armadillo. He also admits that there is a great advantage on the side of the armadillo in the possibility of studying the embryonic conditions.

It is remarkable that the same situations of exact resemblance and of various grades of symmetry reversal occur in both human and armadillo monozygotic twins. The significance of these phenomena must, I believe, be the same in both cases. In the armadillo these manifestations are the result of polyembryonic development; it is almost certainly the same for man.

*Coefficients of correlation between twins.*—The human friction ridges do not furnish as good material as do the bands of the armadillo for establishing an exact numerical measure of the degree of resemblance between twins. An enumeration of the individual friction ridges of the numbers of sweat-pores in a given pattern would give results comparable in availability to the enumeration of scales in the banded region or in individual bands of the armadillo. Doubtless these data could be obtained for human twins, but as yet there is no such information at hand.

Attempts have been made, however, to work out the percentage differences between twins on the basis of a comparison of various physical measurements and weights. Vernon, for example, gives the data for two pairs of identical twins, one of which, aged twenty-three years, showed an average percentage difference for a number of characters of 0.28; while the other, aged twelve, showed a percentage difference of 0.71. Weismann presented the data on one pair of twin brothers, aged seventeen years, which showed a percentage difference of 2.2, nearly ten times that of Vernon's first pair.

Wilder obtained numerous measurements of three pairs of twins, aged respectively 21.10, 17 10, and 17.11 years. They showed an average percentage of difference of. 2.03. Although Wilder realizes that dimensional and other physical measurements are quite unsafe criteria of the genetic resemblance between twins, he is unable to furnish any substitute less objectionable. It is well known that nutritional and other environmental differences greatly affect the size and weight of individuals. In spite of these facts, however, the percentage differences brought out for twins are quite comparable, in so far as they show close resemblance between monozygotic individuals, with the coefficient of correlation brought out for the number of scutes in the bands of the armadillo.

It should be pointed out, however, that fairly marked dimensional *differences*, even at birth, could not be used as evidence against the monozygotic origin of any particular pair of twins; in the armadillo, where the monozygotic origin is specific and unequivocal, there is frequently a striking size difference among the quadruplets of a given set. On the whole, then, it would seem inadvisable to use dimensional measure-

ments either at birth or in later life as data for determin
ing the degree of resemblance (coefficient of correlation)
between twins.

*Modes of inheritance of 'friction-ridge patterns.*—
In his recent studies on the palms and soles of human
beings, Wilder has brought out some very interesting
and significant facts about the inheritance of certain
of these patterns. Without going into the details of
classification or codification of patterns, it may be said
that palm and sole configurations are markedly heritable. A comparison of the prints of the right palm
of a certain father and his six-year-old son[1] show how
exactly these details may be duplicated in two successive
generations. Certain elements in the palm pattern,
for instance, seem to be inherited after the manner of
Mendelian unit characters, strongly marked patterns
of unusual types being dominant over less marked ones.

An interesting possible development from these
facts has to do with the use of these patterns in determining the parentage where there is doubt about the
matter. If a certain unusual type of pattern were
found in either supposed parent and the same pattern
appeared in the supposed offspring, the relationship
might be said to be established with a high degree of
probability. If, on the other hand, such a pattern
failed to appear or appeared only in a highly modified
form, the conclusion of lack of relationship would not
be safe; there is always the chance that a combination
pattern, derived by a mixing of the patterns of the two
parents, might occur, or even that a grandparental
pattern, recessive for one generation, might reappear.

[1] See Wilder, *loc. cit.*

Much more study is needed before this type of evidence could be used as a reliable method of establishing relationships in man.

Nevertheless, it is significant that peculiar patterns in palm- and sole-prints are heritable just as are peculiar patterns in the bands of armor in the armadillo. The various modifications of the inherited pattern are also like the various expressions of doubling in armadillo scutes and bands. The patterns may be reproduced in a more pronounced or in a reduced condition, but they appear to be inherited in the expected proportions on the basis of their unit character nature. Unfortunately scarcely any information has been secured as to the direct inheritance of palm and sole patterns. This field would be well worth investigation.

In concluding the discussion of the friction-ridge correspondences it will be of interest to describe a case worked out for a pair of "conjoined" twins (pygo-pagi) by Wilder. These twin girls (Margaret and Mary) were first observed a day or two after birth and are now over four years old. "They are united in the sacro-iliac region, but are placed somewhat obliquely, so that instead of looking in opposite directions they are rotated about 45 degrees toward the same side." A study of their palms and soles was deemed of great interest, since there appeared to be no question as to their monozygotic derivation. The palms of all four hands are practically alike in pattern; the right hand of each one not only mirrors its own left hand but also the left hand of its twin partner. Since there is no asymmetry in the hands, there is no chance to observe symmetry reversal or mirror-imaging. Strange to say, however, although both sole patterns of Mary and the left one of Margaret are alike (Fig. 55), the right sole print of Margaret had a totally different configuration.

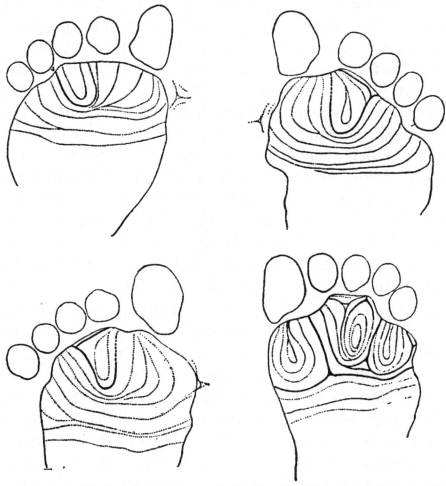

FIG. 55.—Sole-prints of the conjoined twins Margaret (above) and Mary (below). Note that the right of Margaret is a mirror-image of the left of Mary, but that the right of Mary shows a totally different pattern from her left. (From Wilder.)

The left sole of Mary and the right sole of Margaret are more nearly identical than are the right and left soles of Margaret, which is a good case of mirror-imaging and just the kind of thing one might look for in con joined twins. Quite unexpected, however, is the occurrence of the odd pattern in the right sole of Mary.

I am inclined to interpret the cause of this aberrant sole pattern in the light of similar conditions found in armadillo quadruplets. There it was not unusual to find that one or more fetuses in a set inherited a peculiarity while others did not. This was explained as an instance of somatic segregation taking place during the early cleavage divisions. Evidently this is an instance of a similar phenomenon occurring in conjoined twins. In principle this is not different from the unilateral reappearance in an ordinary offspring of a peculiarity inherited from a parent. The case is one of very special interest, since it is the only one of twins on record where the embryonic membranes were studied in correlation with the somatic resemblances. It was ascertained that "there was a single chorion without trace of a separating partition, and the placenta was bilobed, and nearly as large as two normal placentae." The umbilical cord was single for 11 cm. from the placenta, forking into two branches, one a little larger than the other, running to the two individuals. It is only to be regretted that the palm and sole patterns of the two parents were not recorded. Possibly these data, quite crucial in character, I believe, may be yet available; it would probably demonstrate the fact of somatic segregation of parental characters.

*Variations on brain convolutions and in hair arrangement in duplicate twins.*—A significant paper by Sano[1]

[1] F. Sano, *Philosophical Transactions of the Royal Society of London*, CCVIII (1916).

has just appeared in which a detailed comparison of the convolutional pattern of the brains of a pair of stillborn, full-term, Belgian war babies.  These were adjudged, and probably correctly, to be "identical" or mono-zygotic twins, although there appear to be marked dif-ferences in size, weight, facial and cranial indices, size of brain and of head.  "The boy called A has a more receding forehead;  the nose is more turned up;  the distance between the root of the nose and the superior border of the upper lip smaller (6.5 mm. *v.* 8. mm.), hence the mouth remains open.  The chin is more receding.  The ear of A is closer to the head, has very little enrolment of the border, and its lobule is adherent, while the second boy's ear is more unfolded and graceful."

Although in these and other respects the twins are quite strikingly different, they are alike in eyes, in lines and furrows of the hands, and in other inherited char-acters.  The author concludes "that the male twins under examination are very similar to each other and also to their mother.  No essential differences were to be found."

The description of the brains of the twins is very detailed and technical, but we may accept the author's conclusion that, although the brain of twin B is larger and somewhat more advanced in structure than that of twin A, they are strikingly similar.  My own opinion as to the other differences pointed out by Sano is that they too are to be interpreted as the result of a difference in the degree of maturity of the two twins, B being distinctly in advance of A.  It will be recalled that armadillo quadruplets in advanced pregnancy show equally striking differences in developmental age.

The point that interested me most in Sano's paper has to do with the crown whirls of the two twin boys. That of A is to the left of the median line and that of B to the right. In detail the hair whorls of the two are mirror-image duplicates. Such a condition is just what we might expect in monozygotic twins, for there is an intimate relation between hair arrangements, scale arrangements, and friction-skin patterns; they are all integumentary structures and are therefore likely to exhibit mirror-imaging. Were there no other evidence of the monozygotic character of these twins, this condition of the hair whorls would go far to prove it.[1]

That many of the differences between such monozygotic twins may be merely the result of differences in developmental age is shown by an interesting pair of very early human twins that were studied by Dr. F. E. Chidester. One of these twins was about as advanced as a month-old embryo and the other was in an early primitive-streak stage. They both lay in a single large amnion and were separated by a small area of extra-embryonic tissue. Dr. Chidester kindly showed me a drawing of a surface preparation of these twins, but only a detailed study of sections will reveal the interrelations of the two, and I shall be much interested to learn the outcome of this study

*Mental resemblances between duplicate human twins.*— If twins are strikingly alike structurally, it follows that they must be alike functionally, since similar structures could hardly have dissimilar functions. A pair of

[1] I have just learned of an authentic case of reversal in a pair of duplicate twins, one of whom is a colleague of mine here in the University of Chicago. One of these twin brethren is right-handed and the other left-handed.

twins with identical brains should have identical mental equipment at birth, but further mental development would depend on training and environment. Many accounts have been given of the remarkable unity of thought and action of duplicate twins. They have been described as speaking in unison with the same inflection, as having the same dreams, etc. Even more remarkable are stories to the effect that if one twin becomes ill the other does also, and that an injury to one twin is felt by the other. Such anecdotes, though common, are probably without foundation. More credible are stories that one twin got two baths and the other none, or that one twin was spanked for the other's misdeeds.

THE COMPARATIVE POTENCY OF HEREDITY
AND ENVIRONMENT

To what extent and within what limits are the definitive characters of the individual determined at the time of fertilization, and in how far are the minutiae of organic structure to be considered as the product of individual variability beyond the limits of hereditary control? This very fundamental question has been raised under various guises for many years. It is the old problem as to the relative potency of heredity and environment in development, or that of predetermination versus epigenesis.

As long ago as 1870 Galton previsioned the importance of the study of twins as material probably adapted to a solution of the problem. His views are very clearly stated in his paper "The History of Twins, as a Criterion of the Relative Power of Nature and

Nurture." In a subsequent paper (1892) he shows his continued interest in this problem by his remark: "It may be mentioned that I have an enquiry in view which has not yet been fairly begun, namely: to determine the minutest biological unit that may be hereditarily transmissible. The minutiae in the finger-prints of twins seem suitable objects for this purpose."

Wilder in his paper on "Duplicate Twins and Double Monsters" follows up this clue and presents many important facts as to the close resemblance between twins in the patterns of the friction ridges in palms and soles. The conclusion reached is as follows:

The influence of the germ-plasm and its mechanism [i.e., the direct control exercised by heredity] is exerted upon the friction-skin surfaces only so far as concerns the general con-figuration, i.e., the main lines, the patterns, and other similar features; the individual ridges and their details [minutiae] are apparently under the control of individual mechanical laws to which they are subjected during growth. Have we then arrived at the limit of the control of the predetermining mechanism beyond which mechanical laws are alone operative, and is it then possible to hold that the modifications in the latter field are the results of individual experience, and that they are similar in the various members of a given species solely because of similar environment? To these and similar questions we can give no answer at present; yet it seems likely that in general in the palm and sole markings, not only in man but in other mammals as well, we have a set of easily observed and very significant data which may yield important results to future investigators.

Data similar to that on the friction-ridge patterns of human twins are afforded by a study of scute and band doubling in armadillo quadruplets. Just as there is in human twins usually a striking general similarity between the main patterns of the two individ uals and a .difference in the exact detailed expression of the pattern in terms of integumentary units, so in armadillo quadruplets there is generally a pattern (double band or double scute) in a similar region of the armor in all four individuals of a set, but there may be a considerable difference in the number and distribution of the integumentary units employed in the expression of the pattern. The pattern (doubling) may be expressed in a large number of units (scutes) in some members of a quadruplet set and in a small number (sometimes only one) of units in others. It may be expressed unilaterally in some and bilaterally in others. Or, finally, the pattern may be expressed in some members and totally suppressed in others.

When, in his examination of same-sexed twins, Wilder encountered cases in which certain patterns were not sufficiently identical to meet his preconceived ideas of what duplicate twins should be, he concluded that they were fraternal twins (dizygotic). In the light of what I have found in armadillo quadruplets, which are unquestionably monozygotic, it does not seem safe to exclude from the category of monozygotic twins those that fail to show identity of pattern; the same practice, if applied to armadillo quadruplets, would lead to grave error. It is probably safer to say that some monozygotic human twins, like some sets of armadillo quadruplets, are nearly identical, while others, like various quadruplet sets, may differ materially from each other.

To me it appears almost certain that Wilder's twins, Nos. VII, X, and XIII, are monozygotic (duplicates), although they show differences in the presence of certain friction-ridge patterns. It is interesting to note the doubt in Wilder's mind as to the nature of these twins.

Of set VII, he says:

This case has caused me considerable trouble, owing to a preconceived notion that the marks ought to be found identical. The family emphasized the facial resemblance of these twins, and when I first saw them they certainly looked alike. One was, however, an inch taller than the other, and the facial resemblance after a short acquaintance did not seem as great. Upon unprejudiced comparison the prints of the palms are very different, and not at all as in the case of true duplicates. The finger patterns also do not at all correspond. The sole markings are similar but not identical. The case is plainly one of fraternal twins that resemble one another somewhat more than the average.

In this connection let us recall, for a moment, the case of the conjoined twins, Margaret and Mary, described by Wilder in a later paper. In that case the palm-prints were nearly identical, but the right sole of Mary was totally different from her left and from either sole of Margaret. If the criteria employed for set VII were applied to them, the twins Margaret and Mary would be excluded from the category of duplicates; yet there can be no question as to their monozygotic origin.

It must be emphasized also that a difference of one inch in height in fifteen-year-old girls cannot be considered as evidence of dizygotic origin; in armadillo quadruplets there are frequently much greater discrepancies in size than this. It should furthermore not be forgotten that the sole markings are similar in this pair (Wilder's VII) and would doubtless have caused the twins to be classed as identical had not the palm prints been found different.

Of twins No. *X*, Wilder says:

> These twins caused me some little difficulty, although they show by the formulae great differences and determine the set as fraternal beyond a doubt.  The subjects are little girls of ten, whom I have seen but once, and at the time I took it for granted that they were duplicates, and as they came to my laboratory hand-in-hand, dressed exactly alike, and each with her hair in two small braids, they were certainly similar, but to my assistant they did not appeal in the same way, and she judged them fraternal before seeing the prints.  There is a noticeable difference in height and quite a little in weight, greater than is usually found in true duplicates.

Again, it is highly probable that this pair is mono-zygotic, although there is a variation in the distribution of patterns in the two individuals.  Even an examination of finger-prints reveals a very close similarity, the difference being in the palms of the right hands.  Sole-prints are not mentioned.

Of twins No. *XIII*, Wilder says:

> According to personal appearance these should be duplicates. I have never seen them, but the one who took the prints wrote: "The Misses . . . . are so similar in coloring and features that even their best friends confuse them."   It must be confessed, how-ever, that the differences in the formulae cannot be reconciled, and that the palms are, and remain, in respect to the main lines, very different.  They both possess, however, certain peculiar markings in common, as the thenar patterns in the left hands, or the hypothenar convergence in the right hands, facts which would help matters out were there any hope of reconciling the lines.   I must leave this as a totally aberrant case and treat it as such in the summary given below.   In the other two cases that have caused trouble, Nos. VII and X, the resemblance is not so striking

and there are marked differences in height and weight. It will be noted that in these there is a complete lack of the bilateral symmetry in the hands of one individual, which is usual, though not invariably the case, in undoubted duplicates. Were the theory established beyond a doubt, I should unhesitatingly diagnose this as a case of fraternals in whom there happens to be striking resemblance, but as one cannot be dogmatic, I must leave it as recorded, without explanation. The finger prints correspond exactly in the two individuals, even more than is usual in those twins that are unquestionably duplicates, yet it will be noted that they are, in the main, ulnar loops, the commonest type of pattern.

This, in my opinion, is unquestionably a case of duplicates and exhibits conditions quite parallel to those shown by armadillo quadruplets.

These rather long but significant quotations serve to show the difficulty of applying, as a criterion of monozygotic origin of twins, resemblances or lack of resemblances in any unit characters. It would seem, on the whole, more feasible to trust to one's judgment of the general similarity in features, coloring, disposition, and the like, for such resemblances are at least as important elements in the personality as are finger-prints; a general summation of resemblances is more likely to be a sound basis than any single detail could be, especially since we know that monozygotic armadillo quadruplets often differ markedly among themselves in respect to characters of strictly comparable nature. The presence of any type of symmetry reversal would to my mind outweigh any lack of detailed resemblance in deciding that any given set of twins is monozygotic.

Whether in the light of these circumstances Wilder's idea is justifiable—that we can measure the limits of hereditary control in twins by concluding *that the general configurations of friction-skin patterns is predetermined and only the minutiae are beyond the limits of hereditary control*—is a question which will have suggested itself to the reader before this. This conclusion was based on those cases of duplicates which were alike in the general configuration of the patterns; but if, as seems certain, monozygotic twins sometimes differ not only in minutiae but in the general configuration of friction ridges, this conclusion as to the limits of hereditary control fails to hold.

It certainly fails to hold for armadillo quadruplets, which are uniformly monozygotic.

### HEREDITARY CONTROL AND SOMATIC SEGREGATION

To find a failure of bilaterality in certain friction-ridge patterns in a single individual is not at all unusual. For example, when I examine the friction patterns of my own hands, I find a pronounced dissimilarity in the two. The right hand has a very conspicuous hypothenar whorl not even suggested in the left. In the left hand there occurs a well-defined triradius, absent in the right. In other respects the two hands are similar but not identical. If hereditary control is to be tested by a comparison of duplicate twins, which are *believed* to be monozygotic, why would it not be much simpler to test it by a comparison of the antimeric halves of a single individual who is *known* to be monozygotic? Unquestionably, if by hereditary control is meant that identity of two or more homologous or bilateral products of a single zygote is predetermined, even the main patterns of friction ridges cannot be said to be hereditarily controlled in the same way on both sides of the body of those individuals who are bilaterally asymmetrical.

The unilateral appearance of an inherited unit character, such as a friction-skin pattern, almost certainly implies some unilaterality in the somatic distribution of the differentiating factor for this character. Whether the character appears in one or in both of a pair of twins (which are genetically equivalent to the right and left sides of a single individual), or, finally, whether it appears in one, two, three, or four members of a set of armadillo quadruplets, depends on whether the differentiating factor is distributed during the earliest cleavage in a unilateral or bilateral fashion; in other words, whether, with respect to the differentiating factor in question, the earliest cleavages have been equational or differential. All of the cells derived from the blastomere that receives the factor will produce individuals with the character, unless, as often happens, subsequent cleavages still further limit the distribution of the factor by repeated differential division.

Thus we appear to have the possibility of a segregative mechanism, which, in so far as an individual or set of monozygotic twins is concerned, might give results that would resemble the segregation of unit characters in the maturation division of the germ-cells. That segregation of unit characters resulting in the so-called purity of gametes probably has its counterpart in the segregations that occur in the early cleavages in the armadillo and is not confined to the gonads seems certain; there appears to be a parallel segregation in the germ-plasm associated with the soma of each individual. Within a single set of armadillo quadruplets one individual having a unit character (a doubling of integumentary units) in the soma transmits this character to its offspring, while another individual (derived from the same zygote) lacking this character in the soma fails to transmit it to its offspring. In conclusion, I should therefore like to emphasize the fact that somatic divisions may be as important agents in segregating unit characters as germinal divisions involved in the formation of gametes (maturation or reduction divisions) are believed to be. On this theory it might well happen during cleavage that cells derived from one part of a single gonad would, prior to maturation, contain certain determiners which others in another part of the same gonad lack.

It would appear, then, that the limits of hereditary control are not to be measured by a comparison of twins, or even by comparing the antimeric halves of the same individual. The whole question hinges on the equality or inequality of distribution during cleavages of the determinative factors; this involves what we have called somatic segregation.

### STATISTICAL METHODS OF DETERMINING THE LIMITS OF HEREDITARY CONTROL

Another more reliable method of testing the limits of hereditary control than those applied to individual cases is the statistical method which applies to large groups. We can find out how strong on the average is the hereditary control exercised by the predeterminative mechanism of the germ-plasm with respect to certain measurable or enumerable characters. In human twins only dimensional differences have been determined, and these are subject to too large an element of environmental control to be used as a test of heredity.

In the armadillo, however, we find an ideal material for statistical study in the armor scutes in the nine bands of armor and in the armor shields. A determination of the coefficient of correlation for large numbers of sets of quadruplets reveals the fact that on the average this coefficient approaches within about 7 per cent of complete identity. For example, the coefficient of correlation derived from an array of 56 sets of male quadruplets gives a coefficient of 0.9294±0.0057. A similar high coefficient was found for 59 sets of female quadruplets. These figures were derived from a study of the total numbers of scutes in the nine bands. Now, when we compare with this the coefficients of correlation for individual bands, it is found that there is a very decided drop from the near approach to identity seen in the banded region as a whole, for the average coefficient of correlation determined for three sample bands (1, 5, and 9) is about 0.4+. These results may be interpreted as showing that the total number of scutes in a large armor region is rather definitely predetermined, but the alignment of the scutes into rows or bands is a process involving developmental mechanics of a cruder sort which appears to be largely beyond the limits of hereditary control. Here then we would appear to be in possession of facts that should enable us to draw the line between "nature and nurture" or to determine the limits of hereditary control. But, as is usual with statistical results, the averaging up of figures conceals the fundamental facts. Studies in the heredity of armor units reveal not infrequently a situation in which, so far as individual bands are concerned, there is a close approach to identity between mother and offspring; in other bands, however, there may be a pronounced lack of heredity. In some cases, moreover, one offspring in a set is almost identical, band for band, with the mother, while another offspring of the same set shows marked differences from the mother. Once more then we shall have to call upon the mechanism of somatic segregation, which is responsible for the segregation of biparental units, so that one individual of a monozygotic set shows the maternal scute count and another shows a very different scute count, presumably like that of the father.

Hence, although a coefficient of correlation derived by averaging the conditions in a large number of monozygotic sets of offspring may be a valuable measure of the average performance of the species, it has no value when applied to individual cases. One must conclude, therefore, that no definite law is to be posited as to the relative potency of "nature and nurture" or of the predeterminative versus the epigenetic factors of development. Every character evidently has a genetic basis in the zygote, but the exact expression of the character is dependent upon developmental or epigenetic factors that vary in each individual case. There appears no longer to be any point to an attempt to determine the relative potency of predeterminative and epigenetic factors in development.

# INDEXES

# INDEX OF SUBJECTS

Germinal vesicle of armadillo egg, 34.
Germ-layer inversion, 42, 108.

Hereditary control, limits of, 6, 175.
Heredity: in armadillo quad-ruplets, 125–55; in human twins, 155–69; of double bands, 133–45; of double scutes, 145–49; of numbers of scutes, 145–49; versus environment, 169.
Hermaphrodite, 96, 97.
Heterosexual, 2, 100.
Homosexual, 2, 100.

Inner-cell mass, 40.
Inversion of germ layers, 42, 108.

Litomastix, 112.

Marsupial cat (*Dasyurus*), 42.
Mataco (*Tolypeutes*), 83.
Mirror-image (mirror-imaging), 4, 17, 152–55.
Mon-amniotic, 23.
Monozygotic, 3, 6.
Monsters: cyclopic, 5; double, 4, 8.
*Mulita* (*Dasypus hybridus*), 27, 65, 68.
Multiple births, 4, 113.

Ovocyte of armadillo, 33.
Ovogenesis of armadillo, 32–35.

Pairing: exceptions to pairing in armadillo quadruplets, 62; of embryos in same, 58.
Parasite (?) of armadillo egg, 86.
Parthenogenesis in armadillo egg, 30, 36.
Peludo (*Euphractus villosus*), 77–83.
Placenta: the development of in *Dasypus*, 66; discoid of *Dasypus*, 58–60; primitive (Träger)

Sex: determination of, 6, 110, 111, 112; differentiation of, 110, 116, 117; ratios in twins and multiple births, 10, 110, 113, 115.
Somatic segregation, 149–52, 174–77.

# INDEX OF AUTHORS

186

CPSIA information can be obtained
at www.ICGtesting.com
Printed in the USA
LVOW03s1931020717
540137LV00033B/1522/P